Medical Entomology for Students
THIRD EDITION

Arthropod vectors of human diseases, such as malaria, filariasis and typhus, are a continuing threat to human health. Since publication of the first edition, *Medical Entomology for Students* has proved a popular textbook by providing the reader with all the basic information on insects, mites and ticks that affect human health. It examines methods of identification, the biology and ecology of these medically important arthropods and how they can be controlled. Its clear presentation and concise writing style, with extensive illustrations and a glossary of entomological and epidemiological terms, make the book comprehensive and accessible. Each chapter concludes with suggestions for further reading.

Medically important arthropods concern students of subjects as varied as tropical medicine, parasitology, entomology and pest control. This textbook provides essential reading for physicians, nurses, health officials and community health workers, and is recommended for both students and teachers of medical entomology courses.

MIKE W. SERVICE is Emeritus Professor of Medical Entomology at the Liverpool School of Tropical Medicine, UK. He has written over 200 research papers on medical entomology and given advice and training in the field to over 50 countries.

Medical Entomology for Students

Third Edition

Mike W. Service

Emeritus Professor of Medical Entomology,
Liverpool School of Tropical Medicine, Liverpool, UK

CAMBRIDGE UNIVERSITY PRESS
Cambridge, New York, Melbourne, Madrid, Cape Town, Singapore, São Paulo

Cambridge University Press
The Edinburgh Building, Cambridge, CB2 8RU, UK

Published in the United States of America by Cambridge University Press

www.cambridge.org
Information on this title: www.cambridge.org/9780521547758

First edition published by Chapman and Hall 1996
Second edition published 2000
Third edition published 2004
Third printing 2007

Printed in the United Kingdom at the University Press, Cambridge

A catalogue record for this book is available from the British Library

Library of Congress Cataloguing in Publication data
Service, M. W.
Medical entomology for students / Mike W. Service. – 3rd edn.
p. cm.
Includes bibliographical references and index.
ISBN 0 521 54775 X (paperback)
1. Insects as carriers of disease. I. Title.
RA639.5.S47 2004
614.4'32 – dc22 2004040659

ISBN-13 978-0-521-54775-8 paperback

To Wednesday once again

Contents

Contents

Preface to the first edition

This is not intended as a reference book on medical entomology; those interested in such a book should consult *Medical Insects and Arachnids* (1993) edited by R. P. Lane and R. W. Crosskey (Chapman & Hall). The present book is aimed at students, whether they be physicians, nurses, health officials, community health workers or those studying for a masters' degree in parasitology or medical entomology. Its aim is to provide basic information on the recognition, biology and medical importance of arthropods and guidelines for their control. In a teaching book such as this it is always difficult to decide how much detail to include and what to omit; you cannot satisfy everyone. Nevertheless I have attempted to write a book to suit the needs of most students.

The reader should be selective. For example, I hope that most will find all, or most of, the information given in the chapters on fleas, lice, bedbugs, scabies mites and flies relevant to their needs, but I would expect readers to be more selective with some other chapters, such as those on mosquitoes. These insects are undoubtedly the most important arthropod vectors; nevertheless some students may think I have included too much detail on certain aspects for their needs, and if so they can largely disregard such bits. I have also tried to be selective and avoid giving too many references at the end of each chapter, but with some vectors this has not been easy.

May 1995

Preface to the second edition

This new edition remains largely unchanged in respect of area covered from the first edition, and I have generally kept to the same style and format. As stated in the first edition this is a student textbook on medical entomology; those interested more in reference books should consult the publications listed at the end of this Preface. I have revised the text where necessary and rewritten many of the sections on control as these can quickly become outdated. A number of new figures are included, and some previous ones have been redrawn or modified. A few of the older references under the headings 'Further reading' have been omitted and several new ones added. Finally I have added a Glossary, mainly entomological, that I hope will help students to understand better some of the entomological terms commonly used.

February 1999

Preface to the third edition

The philosophy remains the same, to try and present clear and concise accounts of the most relevant information on the identification, life cycles and infections transmitted by arthropods of medical importance. The text has been completely revised and updated to include most recent vector and disease control strategies and new discoveries relating to the epidemiology of disease transmission. In 2000 some mosquito species formerly in the genus *Aedes* were transferred to the new genus *Ochlerotatus*, so that for example *Aedes togoi* is now called *Ochlerotatus togoi*. Although this change will inevitably lead to some confusion, I nevertheless have used it here because the name *Ochlerotatus* will now be found in most of the more recent scientific literature. Not to use this new generic name would create greater confusion! Some figures have been redrawn and new ones added, as have new tables and an appendix.

I have tried to help students during revisionary reading by placing in bold italics words relating to items, whether morphological (e.g. ***antennae***, ***capitulum***) or biological (e.g. ***transovarial***, ***reservoir hosts***), that are important in vector recognition or for understanding the role of vectors in disease transmission.

In addition to the Further reading at the end of each chapter, there is a Select bibliography of some key publications, mostly books on medical entomology, after the Glossary.

As before, readers, whether they be physicians, community health workers, health officials, nurses, or those specializing in medical entomology or parasitology, should be selective when reading the various chapters in this book, and focus on facts and issues that are most relevant to their needs or studies. Good luck!

October 2003

Acknowledgements

I would like to thank the following colleagues and friends, arranged in alphabetical order, for their prompt responses to various questions on their specialist subjects, J. P. T. Boorman, J. H. Bryan, R. W. Crosskey, M. J. R. Hall, R. E. Harbach, C. G. Moore, A. Pont and A. Walker.

I am indebted to the Trustees of the Natural History Museum, London for permission to reproduce Figures 4.2 and 4.4a drawn by R. W. Crosskey, and to the Museum for permission to reproduce, in a modified form, Figures 17.1 (male tick) and 19.3. I am also indebted to Blackwell Scientific Publications, Oxford, for permission to reproduce figures 1.14 and 11.5 drawn by Miss M. A. Johnson.

1

Introduction to mosquitoes (Culicidae)

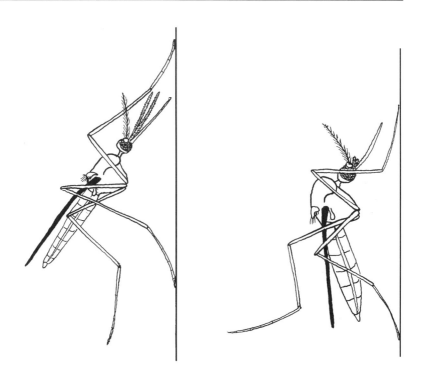

There are some 3300 species of mosquitoes belonging to 41 genera, all contained in the family Culicidae. This family is divided into three subfamilies: Toxorhynchitinae, Anophelinae (anophelines) and Culicinae (culicines). Mosquitoes have a worldwide distribution; they occur throughout the tropical and temperate regions and extend their range northwards into the Arctic Circle. The only areas from which they are absent are Antarctica, and a few islands. They are found at elevations of 5500 m and down mines at depths of 1250 m below sea level.

The most important pest and vector species belong to the genera *Anopheles*, *Culex*, *Aedes*, *Ochlerotatus*, *Psorophora*, *Haemagogus* and *Sabethes*.

Anopheles species, as well as transmitting malaria, are vectors of filariasis (*Wuchereria bancrofti*, *Brugia malayi* and *Brugia timori*) and a few arboviruses. Certain *Culex* species transmit *Wuchereria bancrofti* and a variety of arboviruses. *Aedes* species are important vectors of yellow fever, dengue, encephalitis viruses and many other arboviruses, and in a few restricted areas they are also vectors of *Wuchereria bancrofti* and *Brugia malayi*. Species in the very closely related genus *Ochlerotatus* also transmit filariasis and encephalitis viruses. *Mansonia* species transmit *Brugia malayi* and sometimes *Wuchereria bancrofti* and a few arboviruses. *Haemagogus* and *Sabethes* mosquitoes are vectors of yellow fever and a few other arboviruses in Central and South America, while the genus *Psorophora* contains some troublesome pest species in North and South America, as well as a few transmitting arboviruses.

Many species, although not carriers of any disease, can nevertheless be troublesome because of the serious biting nuisances they cause.

1.1   EXTERNAL MORPHOLOGY OF MOSQUITOES

Mosquitoes possess only one pair of functional wings, the fore-wings. The hind-wings are represented by a pair of small, knob-like halteres. Mosquitoes are distinguished from other flies of a somewhat similar shape and size by: (1) the possession of a conspicuous forward-projecting proboscis; (2) the presence of numerous appressed scales on the thorax, legs, abdomen and wing veins; and (3) a fringe of scales along the posterior margin of the wings.

Mosquitoes are slender and relatively small insects, usually measuring about 3–6 mm in length. Some species, however, can be as small as 2 mm while others may be as long as 19 mm. The body is distinctly divided into a head, thorax and abdomen.

The head has a conspicuous pair of kidney-shaped compound eyes. Between the eyes arises a pair of filamentous and segmented antennae. In females the antennae have whorls of short hairs (that is pilose antennae), but in males, with a few exceptions in genera of no medical importance, the antennae have many long hairs giving them a feathery or plumose appearance. Mosquitoes can thus be conveniently sexed by examination

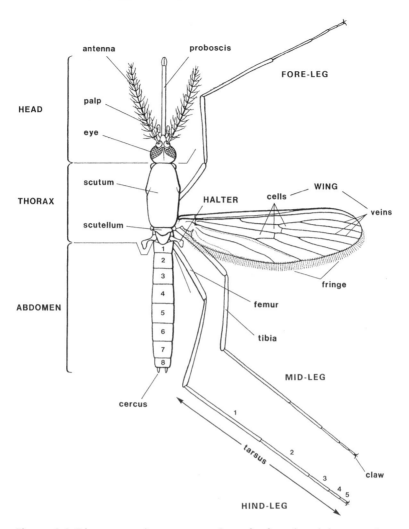

Figure 1.1 Diagrammatic representation of a female adult mosquito.

of the *antennae*: individuals with feathery antennae are males, whereas those with only short and rather inconspicuous antennal hairs are females (Figs. 1.1, 1.13). Just below the antennae is a pair of *palps* which may be long or short and dilated or pointed at their tips, depending on the sex of the adults and whether they are anophelines or culicines (Fig. 1.13). Arising between the palps is the single elongated proboscis, which contains the piercing mouthparts of the mosquito. In mosquitoes the proboscis characteristically projects forwards (Fig. 1.1).

The thorax is covered, dorsally and laterally, with scales which may be dull or shiny, white, brown, black or almost any colour. It is the arrangement of black and white, or coloured, scales on the dorsal surface of the

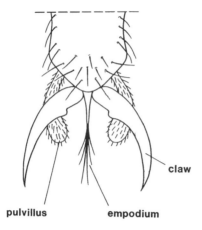

claw

pulvillus   **empodium**

Figure 1.2 Tip of the last segment of the tarsus of a *Culex* mosquito showing claws, hair-like empodium and two large pulvilli.

thorax that gives many species (especially those of the genera *Aedes* and *Ochlerotatus*) distinctive patterns (Fig. 3.3).

The wings are long and relatively narrow, and the number and arrangement of the wing veins is virtually the same for all mosquito species (Fig. 1.1). The veins are covered with scales which are usually brown, black, white or creamy yellow, but more brightly coloured scales may occasionally be present. The shape of the scales and the pattern they form differs considerably between both genera and species of mosquitoes. Scales also project as a fringe along the posterior border of the wings. In life the wings of resting mosquitoes are placed across each other over the abdomen in the fashion of a closed pair of scissors. The legs of mosquitoes are long and slender and are covered with scales which are usually brown, black or white and may be arranged in patterns, often in the form of rings (Fig. 3.4b). The tarsus usually terminates in a pair of toothed or simple claws. Some genera, such as *Culex*, have a pair of small fleshy pulvilli (Fig. 1.2) between the claws.

The abdomen is composed of 10 segments but only the first seven or eight are visible. In mosquitoes of the subfamily Culicinae, the abdomen is usually covered dorsally and ventrally with mostly brown, blackish or whitish scales. In the Anophelinae, however, the abdomen is almost, or entirely, devoid of scales. The last abdominal segment of the female mosquito terminates in a pair of small finger-like cerci, whereas in the males a pair of prominent claspers, comprising part of the male external genitalia, are present.

In unfed mosquitoes the abdomen is thin and slender, but after females have bitten a suitable host and taken a blood-meal (only females bite) the abdomen becomes greatly distended and resembles an oval red balloon.

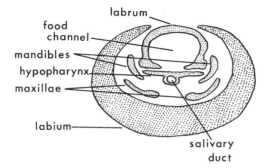

Figure 1.3 Diagram of a cross-section through the proboscis of a mosquito showing components of the mouthparts and food channel.

When the abdomen is full of developing eggs it is also dilated, but is whitish and not red in appearance.

1.1.1 Mouthparts and salivary glands

The mouthparts are collectively known as the ***proboscis***. In mosquitoes the proboscis is elongate and projects conspicuously forwards in both sexes – although males do not bite. The largest component of the mouthparts is the long and flexible gutter-shaped labium which terminates in a pair of small flap-like structures called the labella. In cross-section the labium is seen to almost encircle all the other components of the mouthparts (Fig. 1.3) and serves as a protective sheath. The individual components are held close together in life and only become partially separated during blood-feeding, or when they are teased apart for examination as illustrated in Figure 1.4.

The uppermost structure, the labrum, is slender, pointed and grooved along its ventral surface. In between this 'upper roof' (labrum) and 'lower gutter' (labium) are five needle-like structures, namely, a lower pair of toothed maxillae, an upper pair of mandibles, which usually lack teeth (although in *Anopheles* they are very finely toothed), and finally a single untoothed, hollow stylet called the hypopharynx. When a female mosquito bites a host the labella, at the tip of the fleshy labium, are placed on the skin and the labium, which cannot pierce the skin, curves backwards. This allows the paired mandibles, paired maxillae, labrum and hypopharynx to penetrate the host's skin. Saliva from a pair of trilobed salivary glands, situated ventrally in the anterior part of the thorax, is pumped down the hypopharynx. *Saliva* contains antihaemostatic enzymes, such as apyrase, that produce haematomas in the skin and facilitate the location and uptake of blood. Saliva also contains anticoagulants which prevent the blood from clotting and obstructing the mouthparts as it is sucked up into the space formed by the apposition of the labrum and other piercing mouthparts.

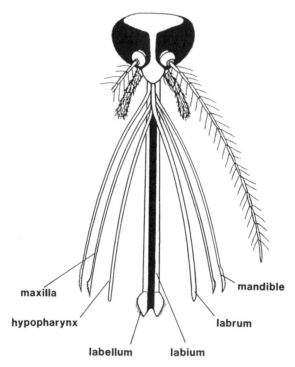

maxilla

hypopharynx

labellum   labium

mandible

labrum

Figure 1.4 Diagram of the head of a female cuclicine mosquito showing the components of the mouthparts spread out from the labium.

Anaesthetic substances help reduce the pain inflicted by the mosquito's bite, so reducing the host's defensive reactions.

Although male mosquitoes have a proboscis, the maxillae and mandibles are usually reduced in size or the mandibles are absent, so males cannot bite.

1.2   LIFE CYCLE

1.2.1   Blood-feeding and the gonotrophic cycle

Most mosquitoes mate shortly after emergence from the pupa. After mating, sperm passed by the male into the female enters her spermatheca. This sperm, in the spermatheca, usually serves to fertilize all eggs laid during her lifetime; thus only one mating and insemination per female is required. With a few exceptions, a female mosquito must bite a host and take a blood-meal to obtain the necessary nutrients for the development of the eggs in the ovaries. This is the normal procedure and is referred to as *anautogenous* development. A few species, however, can develop the first batch of eggs without a blood-meal, and more rarely subsequent batches. This process is called *autogenous* development. The speed of digestion of

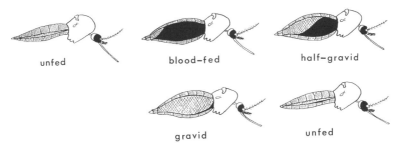

unfed blood-fed half-gravid

gravid unfed

Figure 1.5 Diagrammatic representation of the gonotrophic cycle of a female mosquito. Each cycle begins with an unfed adult which passes through a blood-fed, half-gravid and gravid condition. After oviposition the female is again unfed and seeks another blood-meal.

the blood-meal depends on temperature. In most tropical species it takes only 2–3 days, but in colder, temperate countries blood digestion may take as long as 7–14 days.

After a blood-meal the mosquito's abdomen is dilated and bright red in colour, but some hours later the abdomen becomes a much darker red. As the blood is digested and the white eggs in the ovaries enlarge, the abdomen becomes whitish posteriorly and dark reddish anteriorly. This condition represents a mid-point in blood digestion and ovarian development, and the mosquito is referred to as being half-gravid (Fig. 1.5). Eventually all blood is digested and the abdomen becomes dilated and whitish due to the formation of fully developed eggs (Fig. 1.5). The female is now said to be gravid and she searches for suitable larval habitats in which to lay her eggs. After oviposition the female mosquito takes another blood-meal and after 2–3 days (in the tropics) a further batch of eggs is matured. This process of blood-feeding and egg maturation, followed by oviposition, is repeated several times throughout the female's life and is referred to as the *gonotrophic cycle*.

Male mosquitoes cannot bite but instead feed on the nectar of flowers and other naturally occurring sugary secretions. Males are consequently unable to transmit any diseases. Sugar feeding is not, however, restricted to males: females may also feed on sugary substances to obtain energy for flight and dispersal, but only in a few species (the autogenous ones) is this type of food sufficient for egg development.

1.2.2 Oviposition and biology of the eggs
Depending on the species, female mosquitoes lay about 30–300 eggs at any one oviposition. Eggs are brown or blackish and 1 mm or less long. In many Culicinae they are elongate or approximately ovoid in shape, but eggs of *Mansonia* are drawn out into a terminal filament (Fig. 3.8). In the Anophelinae eggs are usually boat-shaped (Fig. 1.8). Many mosquitoes, such as

species of *Anopheles* and *Culex*, lay their eggs directly on the water surface. In *Anopheles* the eggs are laid singly and float on the water, whereas those of *Culex* are laid vertically in several rows held together by surface tension to form an egg raft which floats on the water (Fig. 1.15). *Mansonia* species lay their eggs in a sticky mass that is glued to the underside of floating plants. None of the eggs of these mosquitoes can survive desiccation and consequently they die if they become dry. In the tropics eggs hatch within 2–3 days, but in cooler temperate countries they may not hatch until after 7–14 days, or longer.

Other mosquitoes, such as those belonging to the genera *Aedes, Ochlerotatus, Psorophora* and *Haemagogus*, do not lay eggs on the water surface but deposit them just above the water line on damp substrates, such as mud and leaf litter, or on the inside walls of tree-holes and clay water-storage pots. Eggs of these genera can withstand desiccation, especially those of *Aedes, Ochlerotatus* and *Psorophora* which can remain dry for months or even years but still remain viable and hatch when soaked in water. Because such eggs are laid above the water line of breeding places it may be many weeks or months before they become flooded with water, and thus have the opportunity to hatch. However, even when flooded, hatching may extend over long periods because the eggs hatch in instalments. Moreover, eggs of *Aedes, Ochlerotatus* and *Psorophora* may require repeated immersions in water followed by short periods of desiccation before they will hatch. *Aedes, Ochlerotatus* and *Psorophora* eggs may also enter a state of either quiescence, only hatching when suitable conditions occur, or ***diapause***, not hatching until some specific stimulus terminates the state of diapause. Environmental stimuli such as changes in daylength and/or temperature often break diapause and cause eggs to hatch. In temperate regions many *Aedes, Ochlerotatus* and *Psorphora* species overwinter as diapausing eggs.

1.2.3   Larval biology

Mosquito larvae can be distinguished from other aquatic insects by being legless and having a bulbous thorax that is wider than both the head and the abdomen. There are four active larval instars. All mosquito larvae require water in which to develop; no mosquito has larvae that can withstand desiccation although they may be able to survive short periods, for example, in wet mud.

Mosquito larvae have a well-developed head, bearing a pair of antennae and a pair of compound eyes. Prominent mouthbrushes are present in most species and serve to sweep water containing minute food particles into the mouth. The thorax is roundish in outline and has various simple and branched hairs which are usually long and conspicuous. The 10-segmented abdomen has nine visible segments, most of which have simple or branched hairs (Figs. 1.9, 1.16). The last segment, which differs in shape from the preceding eight segments, has two paired groups of long hairs forming the caudal setae, and a larger fan-like group comprising the *ventral brush* (Figs. 1.10, 1.16). This last segment ends in two pairs of transparent, sausage-shaped anal papillae, which although often called gills are not concerned with respiration but with osmoregulation.

Mosquito larvae, with the exception of *Mansonia* and *Coquillettidia* species (and a few other species), must come to the water surface to breathe. Atmospheric air is taken in through a pair of spiracles situated dorsally on the tenth abdominal segment. In the subfamilies Toxorhynchitinae and Culicinae these spiracles are situated at the end of a single dark-coloured and heavily sclerotized tube termed the **siphon** (Fig. 1.16). *Mansonia* and *Coquillettidia* larvae possess a specialized siphon that is more or less conical, pointed at the tip and supplied with prehensile hairs and serrated cutting structures (Fig. 3.9). These enable the siphon to be inserted into the roots or stems of aquatic plants and thus oxygen for larval respiration is obtained from the plants. In contrast larvae of the Anophelinae do not have a siphon (Figs. 1.10, 1.13).

Mosquito larvae feed on yeasts, bacteria, protozoans and numerous other micro-organisms, as well as on decaying plant and animal material found in the water. Some, such as *Anopheles* species, are surface-feeders, whereas many others browse over the bottoms of habitats. A few mosquitoes are carnivorous or cannibalistic. There are four larval instars and in tropical countries larval development, that is the time from egg hatching to pupation, can be as short as 5–7 days, but many species require about 7–14 days. In temperate areas the larval period may last several weeks or months, and several species overwinter as larvae.

1.2.4   Larval habitats

Mosquito larval habitats vary from large and usually permanent collections of water, such as freshwater swamps, marshes, ricefields and borrow pits, to smaller collections of temporary water such as pools, puddles, water-filled car tracks, ditches, drains and gulleys. A variety of 'natural container-habitats' also provide breeding places, such as water-filled tree-holes, rock-pools, water-filled bamboo stumps, bromeliads, pitcher plants, leaf axils in banana, pineapple and other plants, water-filled split coconut husks and snail shells. Larvae also occur in wells and 'man-made container-habitats', such as clay pots, water storage jars, tin cans, discarded kitchen utensils and motor vehicle tyres. Some species prefer shaded larval habitats whereas others like sunlit habitats. Many species cannot survive in water polluted with organic debris whereas others can breed prolifically in water contaminated with excreta or rotting vegetation. A few mosquitoes breed almost exclusively in brackish or salt waters, such as saltwater marshes and mangrove swamps, and are consequently restricted to mostly coastal areas. Some species are less specific in their requirements and can tolerate a wide range of different types of breeding place.

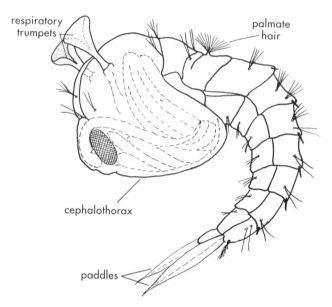

respiratory
trumpets

palmate
hair

cephalothorax

paddles

Figure 1.6 *Anopheles* pupa.

Almost any collection of permanent or temporary water can constitute a
mosquito larval habitat, but larvae are usually absent from large expanses
of uninterrupted water such as lakes, especially if they have large numbers
of fish and other predators. They are also usually absent from large rivers
and fast-flowing waters, except that they may occur in marshy areas and
isolated pools and puddles formed at the edges of flowing water.

1.2.5 Pupal biology

All mosquito pupae are aquatic and comma-shaped. The head and thorax
are combined to form the cephalothorax, which has a pair of respiratory
trumpets dorsally (Fig. 1.6). The abdomen is 10-segmented although only
eight segments are visible. Each segment has numerous short hairs and the
last segment terminates in a pair of oval and flattened structures termed
paddles (Figs. 1.11, 1.18). Some of the developing structures of the adult
mosquito can be seen through the integument of the cephalothorax, the
most conspicuous features being a pair of dark compound eyes, folded
wings, legs and the proboscis (Fig. 1.6).

Pupae do not feed but spend most of their time at the water surface
taking in air through the respiratory trumpets. If disturbed they swim up
and down in the water in a jerky fashion.

Pupae of *Mansonia* and *Coquillettidia* differ in that they have relatively
long breathing trumpets, which are modified to enable them to pierce
aquatic vegetation and obtain their oxygen in a similar fashion to the larvae
(Fig. 3.9). As a consequence their pupae remain submerged and rarely come
to the water surface.

In the tropics the pupal period in mosquitoes lasts only 2–3 days but in cooler temperate regions pupal development may be extended over 9–12 days, or longer. At the end of pupal life the skin on the dorsal surface of the cephalothorax splits, and the adult mosquito struggles out.

1.2.6   Adult biology and behaviour

As already mentioned (p. 6), females of most species of mosquito require a blood-meal before the eggs can develop, and this is taken either before or more usually after mating. Many species bite humans to obtain their blood-meals and a few feed on humans in preference to other animals. However, others prefer feeding on non-human hosts and many species never bite people. Species that usually feed on humans are said to be *anthropophagic* in their feeding habits, whereas those that feed mainly on other animals are called *zoophagic*. Mosquitoes that feed on birds are sometimes called ornithophagic instead of zoophagic. Females are attracted to hosts by various stimuli emanating from their breath or sweat, such as carbon dioxide, lactic acid, octenol, as well as body odours and warmth. Vision usually plays only a minor role in host orientation but some species are attracted by the silhouette or movement of their hosts. Some species feed more or less indiscriminately at any time of the day or night; others are mainly diurnal or nocturnal in their biting habits.

A few species of mosquitoes frequently enter houses to feed and are said to be *endophagic* in their feeding habits, whereas those that bite their hosts outside houses are called *exophagic*. After having bitten humans, or some other hosts, either inside or outside houses, mosquitoes seek resting places in which to shelter during digestion of their blood-meals. Some species rest inside houses during the time required for blood digestion and maturation of the ovaries and are called *endophilic*. In contrast mosquitoes that rest outdoors are termed *exophilic*. Female adults of *Aedes aegypti* (a vector of yellow fever), for example, are usually anthropophagic, exophagic and exophilic, whereas adults of *Anopheles gambiae* (African malaria vector) are mainly anthropophagic, endophagic and endophilic. Few mosquitoes, however, are entirely anthropophagic or zoophagic, endophagic or exophagic, endophilic or exophilic. Instead, most show various degrees of these behavioural patterns; in other words all these terms are relative. The feeding behaviour of a species may also change: for example in certain areas and at certain seasons a species may bite people predominantly (anthropophagic) inside houses (endophagic) and remain in houses afterwards (endophilic), whereas at other times, especially if there are few people but many animals in the area, the species may become predominantly zoophagic, and also exophagic and exophilic. Some species are less adaptable in their feeding behaviour and will never rest inside houses or enter them to feed on people.

The *biting behaviour* of female mosquitoes may be very important in the epidemiology of disease transmission. Mosquitoes that feed on people predominantly out of doors and late at night will not bite many young children, because they will be indoors and asleep at this time. Thus, young children will be less likely to be infected with any diseases that these mosquitoes might transmit. During hot and dry periods of the year substantial numbers of people may sleep out of doors and as a consequence be bitten more frequently by exophagic mosquitoes. Some mosquitoes bite predominantly within forests or wooded areas. Consequently, humans will only get bitten when they visit these places. Clearly the behaviour of both people and mosquitoes may be relevant in diseases transmission.

The resting behaviour of adult mosquitoes may be an important consideration in planning control measures. In several malaria control campaigns the interior surfaces of houses, such as walls and ceilings, are sprayed with residual insecticides, such as malathion, to kill adult mosquitoes resting on them. This approach will, of course, only be effective in controlling malaria if the mosquito vectors are endophilic.

Most mosquitoes probably disperse only a few hundred metres from their emergence sites, and in control programmes and epidemiological studies it is usually safe to say that mosquitoes will not fly further than 2 km. There are, however, records of mosquitoes being found up to 100 km or more from their breeding places, but such dispersal is nearly always wind-assisted. Mosquitoes may get transported long distances in aeroplanes, and sometimes this causes disease outbreaks, such as 'airport malaria'.

In tropical countries adult female mosquitoes probably live *on average* 1–2 weeks, whereas in temperate countries adult longevity is likely to be 3–4 weeks. However, some females must live at least two weeks otherwise they would not be vectors of diseases such as malaria. Species that hibernate or aestivate live much longer: for example in Europe some fertilized females of *Culex pipiens* survive in hibernation from August until May. Male adults usually have a shorter life span than females.

1.3   CLASSIFICATION OF MOSQUITOES

1.3.1   Subfamily Toxorhynchitinae

Subfamily Toxorhynchitinae comprises a single genus, *Toxorhynchites*, which contains 76 species that are mainly tropical, although a few species occur in North America, south-eastern Russia and Japan.

Adults are large (19 mm long, 24 mm wingspan) and colourful, being metallic bluish or greenish with black, white or red tufts of hair-like scales projecting from the posterior abdominal segments. Adults are easily recognized by the possession of a proboscis that is curved backwards in both sexes (Fig. 1.7) and is incapable of piercing the skin. Consequently, since neither sex can bite, they are of no medical importance. Their larvae are also large (12–18 mm long), often dark reddish and, like those of the Culicinae,

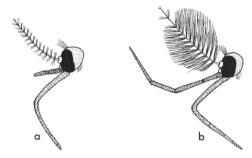

Figure 1.7 Heads of *Toxorhynchites* adults: (a) female; (b) male.

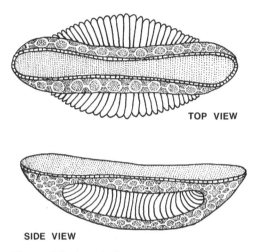

Figure 1.8 *Anopheles* eggs.

have a siphon. They are predacious on larvae of other mosquitoes and on their own kind. They have occasionally been introduced into areas in the hope that their voracious larvae will help reduce the numbers of pest mosquitoes. Larvae are found mainly in container-habitats, such as tree-holes and bamboo stumps, tin cans and water storage pots.

1.3.2 Subfamily Anophelinae
Of the three genera included in the subfamily Anophelinae only the genus *Anopheles* (about 430 species), which contains important malaria vectors, is of medical importance. The following characters serve to separate *Anopheles* mosquitoes from those in any of the other mosquito genera.

Anopheline eggs
Eggs are laid singly on the water surface. In most species they are typically boat-shaped and, laterally, have a pair of air-filled sacs called *floats* (Fig. 1.8). Anopheline eggs are unable to withstand desiccation.

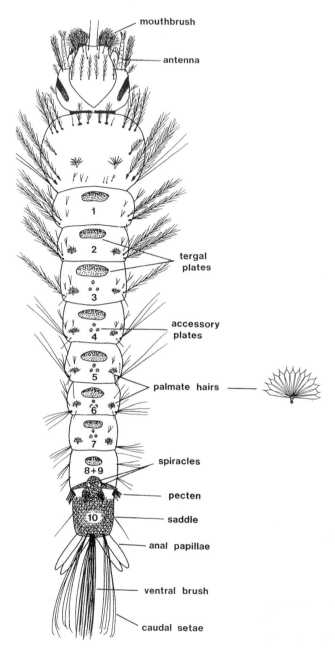

Figure 1.9 *Anopheles* larva, dorsal view, showing diagnostic abdominal tergal plates and palmate hairs.

Anopheline larvae

Larvae lack a siphon and lie *parallel* to the water surface, not subtended at an angle as are the culicines. They are surface-feeders and so spend most of their time at the water surface. Examination under a microscope shows that the abdomen has small, brown, sclerotized plates, called *tergal plates*, on the dorsal surface of abdominal tergites 1–8; in addition there may be 1–3 small accessory plates behind the main tergal plate. In addition most or all of these segments have a pair of well-developed *palmate hairs*, sometimes called float hairs (Figs. 1.9, 1.10). These abdominal palmate hairs and a single pair on the thorax come into contact with the water surface and aid in keeping the larvae parallel to the surface. Laterally on each side of segment 8 (8 and 9 are combined) there is a sclerotized structure with teeth called the *pecten*. All these structures identify larvae as belonging to the genus *Anopheles*.

Anopheline pupae

The respiratory trumpets of anopheline pupae are short and broad distally, thus appearing conical (Figs. 1.6, 1.11a), whereas in most culicines the trumpets are narrower and more cylindrical. The most reliable character for identifying anopheline pupae is the presence of short, peg-like spines situated laterally near the distal margins of abdominal segments 2–7 or 3–7 (Fig. 1.11b); in culicines there are no such spines.

Anopheline adults

Adult *Anopheles* usually rest with their bodies at an *angle* to the surface, that is with the proboscis and abdomen in a straight line (Fig. 1.13). In some species they rest at almost right angles to the surface, whereas in others such as the Indian mosquito *Anopheles culicifacies*, the angle is much smaller. This is a very useful character allowing adults resting in houses and elsewhere to be readily identified as *Anopheles*.

Most, but not all, *Anopheles* mosquitoes have the dark (usually blackish) and pale (usually whitish or creamy white) scales on the wing veins arranged in '*blocks*' or specific areas (Fig. 1.12) forming a distinctive spotted pattern which differs according to species. A few species, however, such as the European *Anopheles claviger*, have the veins covered more or less uniformly with dark (often brown) scales.

The most reliable way to distinguish between adult *Anopheles* and Culicinae is by examination of their *heads*. The first procedure is to determine the sex of the adults: female mosquitoes have non-plumose antennae whereas males have plumose antennae. If the adults are females and also *Anopheles* then the *palps* will be about as long as the proboscis and usually lie closely alongside it (Fig. 1.13). The palps are usually blackish with broad or narrow rings of pale scales, especially on the apical half. In male *Anopheles* the *palps* are also about as long as the proboscis but are distinctly swollen at

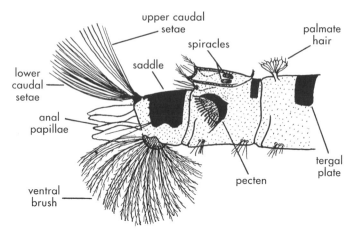

Figure 1.10 Lateral view of the abdominal terminal segments of an *Anopheles* larva.

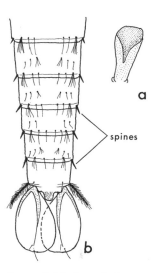

Figure 1.11 *Anopheles* pupa: (a) short and broad respiratory trumpet; (b) part of abdomen showing diagnostic spines.

Figure 1.12 *Anopheles* wing showing dark and pale scales arranged in 'blocks'.

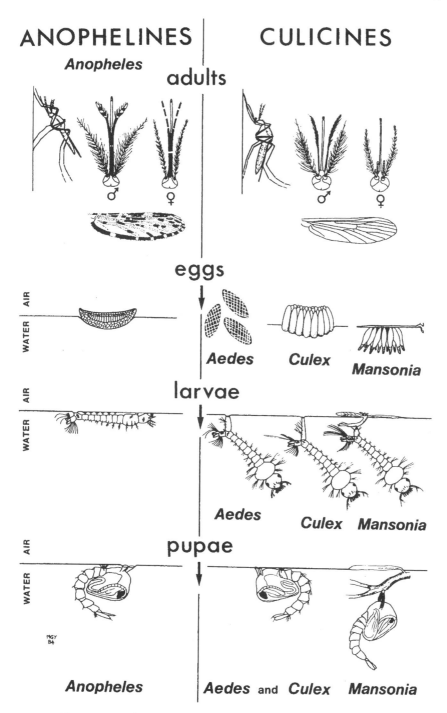

Figure 1.13 Chart of the principal characters of the stages in the life cycle that distinguish anopheline from culicine mosquitoes.

a   b

Figure 1.14 Salivary glands of adult mosquitoes: (a) culicine; (b) *Anopheles* (courtesy of Miss M. A. Johnson; and Blackwell Scientific Publications, Oxford, publishers of *Entomology for Students of Medicine* (1962) by R. M. Gordon and M. M. J. Lavoipierre).

the ends and are said to be clubbed (Fig. 1.13); they may also have rings of pale scales apically.

Other minor differences between *Anopheles* and the Culicinae are that (1) in both sexes of *Anopheles* the scutellum is rounded posteriorly and has setae along the entire edge; (2) only one spermatheca is present in the females; and (3) in both sexes the middle lobe of the salivary glands is considerably shorter than the two outer lobes (Fig. 1.14).

The principal characters for separating the various stages in the life cycles of anopheline and culicine mosquitoes are given in Table 1.1.

1.3.3   Subfamily Culicinae
There are some 2700 species in the large subfamily Culicinae, belonging to 37 genera. The most important medically are the genera *Aedes*, *Ochlerotatus*, *Culex*, *Mansonia*, *Haemagogus*, *Sabethes* and *Psorophora*. The following characters separate the Culicinae from *Anopheles* mosquitoes. Methods for distinguishing between the more important genera within the Culicinae are given in Chapter 3.

Culicine eggs
Culicine eggs never have floats. They are laid either singly (e.g. *Aedes* and *Ochlerotatus*) or in the form of *egg rafts* that float on the water surface (e.g. *Culex* and *Coquillettidia*), or are deposited as sticky masses glued to the underside of floating vegetation (e.g. *Mansonia*) (Fig. 1.15).

Culicine larvae
All culicine larvae possess a *siphon* (Fig. 1.16), which may be long or short. They hang upside down at an angle from the water surface when they are getting air (Fig. 1.13), except for *Mansonia* and *Coquillettidia* larvae which insert their specialized siphons into aquatic plants and remain submerged (Fig. 3.9). There are no abdominal palmate hairs or tergal plates on culicine larvae.

Table 1.1. *Principal characters distinguishing anopheline and culicine mosquitoes*

Stage	Anophelinae	Culicinae
Eggs	Laid singly, possess floats	Laid singly or in egg rafts or masses. Never possess floats
Larvae	Never have a siphon. Lie parallel to water surface. Have abdominal palmate hairs and tergal plates	All larvae have a short or long siphon. Subtend an angle from the water surface. No palmate hairs or tergal plates
Pupae	Breathing trumpets short and broad apically. Short peg-like spines on abdominal segments 2–7 or 3–7	Breathing trumpets short or long, opening not broad. No spines on abdominal segments 2–7
Adults (both sexes)	Rest at an angle to any surface. In most species dark and pale scales on wing veins arranged in distinct 'blocks'	Rest with body more or less parallel to the surface. Scales on wing veins not arranged in 'blocks'; scales frequently all brown or blackish, or a mixture of pale and dark scales scattered on veins
Adult females (non-plumose antennae)	Palps about as long as proboscis	Palps much shorter than proboscis
Adult males (plumose antennae)	Palps about as long as proboscis and swollen at ends	Palps about as long as proboscis but never swollen at ends; palps may be hairy distally

Culicine pupae
The length of the respiratory trumpets in culicine pupae is variable, but they are generally longer, more cylindrical and have narrower openings (Fig. 1.17) than in *Anopheles*. Abdominal segments 2–7 lack peg-like spines although they have numerous setae (Fig. 1.18).

Culicine adults
Culicine adults rest with the thorax and abdomen more or less **parallel** to the surface (Fig. 1.13). The scales covering the wing veins are commonly

Culex   *Mansonia*

Aedes and *Ochlerotatus*

Figure 1.15 Mosquito eggs. *Culex* egg raft that floats on the water surface, *Mansonia* eggs glued to the undersurface of floating aquatic vegetation, and individual *Aedes* or *Ochlerotatus* eggs which are deposited on damp surfaces.

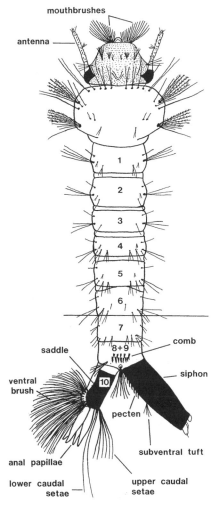

Figure 1.16 Culicine larva, dorsal view but with abdominal segments 7–9 turned laterally to display important identification features.

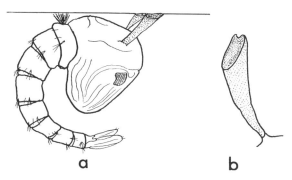

Figure 1.17 Culicine pupa: (a) pupal position at the water surface; (b) one of the two elongated and relatively narrow respiratory trumpets.

Figure 1.18 Part of the abdomen of a culicine pupa showing hair-like setae (note absence of spines as found on anopheline pupae).

uniformly brown or black. Sometimes, however, there are contrasting dark and pale scales but they are not arranged in distinctive areas or 'blocks', as found in many *Anopheles* adults.

The most reliable method for identifying the Culicinae is to examine their **heads**. In females (which have non-plumose antennae) the **palps** are shorter than the proboscis. In males (which have plumose antennae) the **palps** are about as long as the proboscis but are not swollen distally and hence do not appear clubbed (Fig. 1.13). However, the palps may be turned upwards distally, and in many species they are covered with long hairs so that superficially they can appear to be somewhat swollen apically, but more careful examination shows that the palps in male culicines are not clubbed.

Other minor differences that separate the Culicinae from *Anopheles* are that in culicines the scutellum is trilobed and the scutellar setae are restricted to these lobes, whereas in anophelines the scutum is evenly rounded posteriorly and has setae along its entire edge. In culicines there are two or three spermathecae in the females but in anophelines just one, and in culicines the middle lobe of the salivary glands is about as long as the other two, whereas in anophelines it is shorter (Fig. 1.14).

1.4   MEDICAL IMPORTANCE

Although in many temperate countries mosquitoes may be of little or no importance as disease vectors they can, nevertheless, cause considerable annoyance because of their bites. The greatest numbers of mosquitoes are found in the northern areas of the temperate regions, especially near or within the Arctic Circle, where the numbers biting can be so great at certain times of the year as to make almost any outdoor activity impossible. Because of their elongated mouthparts female mosquitoes have little difficulty in biting through clothing such as socks, shirts, blouses, trousers and woollen garments, but clothing with a much closer weave of material may prevent this.

Mosquitoes are important as vectors of malaria, various forms of filariasis and numerous arboviruses, the best known being dengue and yellow fever. Their role in the transmission of these diseases is discussed in Chapters 2 and 3.

1.5   MOSQUITO CONTROL

Greater efforts have been made to control mosquitoes than any other biting insects and a vast literature has accumulated on control operations.

Control measures which are directed against specific vectors, such as anopheline malaria vectors, and *Aedes aegypti* and *Culex quinquefasciatus*, are described in more detail in Chapters 2 and 3; only the more basic principles of control are outlined here.

Control measures can be directed at either the immature aquatic stages or the adults, or at both stages simultaneously.

1.5.1   Control directed at the immature stages
Biological control
Although often termed naturalistic control there is little, if anything, natural about the process of biological control. Either the incidence of predators, parasites or pathogens in habitats must be greatly increased to obtain worthwhile control, or they have to be introduced into habitats from which they were originally absent; such environmental manipulations are not natural. Biological control of mosquitoes was very popular during the early part of the twentieth century, but with the development and availability of chemicals such as the organochlorines and organophosphates it was replaced by insecticidal control. However, because of problems with insecticide resistance and greater awareness of environmental contamination there has been renewed interest in biological (biocontrol) methods. They

are, however, usually more difficult to implement and maintain than insecticidal methods. Moreover, if predators are used it is unlikely that they will prey exclusively on mosquito larvae and pupae; they will also eat harmless or even beneficial insects. Finally, biological control does not lead to rapid control: it takes some days, or more likely weeks, before pest or vector mosquito populations are reduced in size.

Predators   Larvivorous fish are the most widely used biological control agents, the most common being the top minnow or mosquitofish (*Gambusia affinis*). This is a warm-water fish originally native to the southern USA but which has been introduced to over 60 countries worldwide, including the Pacific islands, Europe, the Middle East, India, South-east Asia and Africa, in attempts to control mosquito larvae. They are aggressive fish which have sometimes destroyed indigenous fish; consequently they should not be introduced into new areas. Another commonly used fish is the South American guppy (*Poecilia reticulata*), which is not so voracious as *G. affinis* but can tolerate organic pollution better and is more heat-tolerant. There are numerous other types of fish that have been used to eat mosquito larvae, such as carp (e.g. *Cyprinus carpio*) in Chinese ricefields, an edible catfish (*Clarias fuscus*) in water-storage tanks in Myanmar to control *Aedes aegypti*, and *Oreochromis* (= *Tilapia*) species in Africa and *Aplocheilus* species in Europe and Asia.

Some predatory fish, such as *Aphanius dispar* and *Fundulus* species, breed in saline waters and can therefore be introduced into saltwater habitats. Fish are unsuitable for the control of mosquitoes which breed in small containers, pools and puddles that rapidly dry out. However, some fish, such as species of *Nothobranchius* and *Cynolebias*, which are the so-called instant or annual fish, have drought-resistant eggs and these are more suitable for introducing into small temporary habitats that repeatedly dry out.

Although fish have sometimes greatly reduced the numbers of larvae in certain habitats, such as borrow pits, ponds and ricefields, they have rarely proved effective in reducing the size of mosquito populations over large areas. Nor is there usually much convincing evidence that they have significantly decreased the incidence of mosquito-borne diseases, although in a few mainly arid areas in Iran, Afghanistan, Somalia, Ethiopia, Greece, Russia and China they have been credited with reducing malaria transmission.

Other predators of mosquito larvae and pupae include tadpoles of frogs and toads and various aquatic insect larvae, but these have rarely proved effective as control agents. A few mosquitoes have predacious larvae, for example *Toxorhynchites* species. These have been introduced into container-habitats in certain areas (e.g. Fiji, Samoa and Hawaii) to control larvae of other container-breeding mosquitoes but results have not been very encouraging.

Pathogens and parasites   There are numerous pathogens, such as viruses (e.g. iridescent and cytoplasmic polyhedrosis viruses), bacteria (e.g *Bacillus thuringiensis* var. *israelensis* (= *B. thuringiensis* serotype H-14) and *B. sphaericus*), protozoans (e.g. *Bracheola* (= *Nosema*) *algerae* and *Vavraia culicis*) and fungi (e.g. species of *Coelomomyces*, *Lagenidium* and *Culicinomyces*) that cause larval mortality. There are also several parasitic nematodes that kill mosquito larvae, the best known of which is *Romanomermis culicivorax*, which at one time was commercially mass-produced, but because of non-viable sales is no longer produced. These parasites and pathogens appear harmless to humans.

Although there is considerable interest in biological agents, with the exception of *B. thuringiensis* var. *israelensis* (B.t.i.) *and B. sphaericus*, they cannot yet be routinely advocated for control, especially in tropical countries. *Bacillus thuringiensis* var. *israelensis* is undoubtedly the most useful pathogen as it can be easily mass-produced, is toxicologically safe to humans and wildlife, and is more or less specific in killing mosquito larvae (but it can also kill simuliids: see Chapter 4). It is commonly formulated as a powder that after mixing with water is sprayed on breeding places, but because there is no multiplication of the bacteria there have to be repeated applications, as with most chemical larvicides. However, when formulated as slow-release granules or briquettes, insecticidal activity extends over many days, sometimes up to a month. When B.t.i. is ingested mortality is caused by an endotoxin acting as a stomach poison. Genetic engineering using both non-recombinant and recombinant techniques is improving the larvicidal activity of the bacillus and in addition has transferred the genes responsible for production of the poisonous endotoxin into other bacteria.

Bacillus sphaericus can be formulated much as for *B.t.i.* and kills mosquito larvae in a similar fashion, but differs in that in some situations it can recycle in larval habitats. This species is also more effective in organically polluted waters and is especially effective against *Culex* species. Both these *Bacillus* species are more of a microbial insecticide than a true biological (living) agent that recycles and maintains itself in the environment. Although resistance to both *B.t.i.* and *B. sphaericus* has been observed in laboratory colonies of a few mosquito species, presently this does not pose operational problems, but serves as a warning that control difficulties may arise in the future with both *B.t.i.* and *B. sphaericus*.

Genetic control

Although genetic control methods are directed against the adults rather than the immature stages it is convenient to discuss this strategy here because it is really a form of biological control.

The most common method of genetic control involves the release in the field of sterile male mosquitoes that have been reared in the laboratory. Sterilization can be achieved by ionizing radiation, crossing closely related species to produce infertile hybrid males, or more usually by introducing chemosterilants into larval breeding which makes the emerging adults (both sexes) sterile. The aim is to introduce large numbers of healthy, but sterile, insectary-reared males into field populations that will compete with natural fertile males for female mates, so resulting in large numbers of infertile inseminations. The eggs layed by such females will be sterile and fail to hatch, thus causing a reduction in, or even elimination of, the vector. In El Salvador the release in the 1970s of some 4.36 million chemosterilized male *Anopheles albimanus* (an important malaria vector that had developed resistance to most insecticides) over 4.5 months in an isolated coastal region of about 15 km^2 caused a more than 97% reduction in the biting population. However, because enormous numbers of mosquitoes had to be reared to obtain control over a very limited area, the later expanded trial failed. There is presently little interest in this technique for controlling mosquitoes.

Other methods of genetic control involve the use of cytoplasmic incompatibility, translocations, introduction of lethal genes or genes that make mosquitoes refractory as a vector, and meiotic drive leading to the production of excessive numbers of males. Scientists skilled in genetic engineering are trying to create in the laboratory transgenic mosquitoes that are either incapable of transmitting parasites, such as malaria parasites, or genes that switch vectors from feeding on people to animals. Mechanisms must then be found to drive these genes through the wild population when the transgenic mosquitoes are released in field situations. However, none of these genetic approaches is simple and they will be more difficult to implement than conventional insecticidal methods.

Physical control
Filling-in, source reduction and drainage
Filling-in, source reduction and drainage   This is sometimes referred to as mechanical or environmental control. A simple form consists of filling in and thus completely eliminating breeding places. Larval habitats ranging in size from water-filled tree-holes to ponds and small marshes can be filled in with rubble, earth or sand. Filling in tree-holes can be difficult because many can be at great heights and difficult to locate, or there could be just too many for this method to be practical. Various container-habitats such as abandoned tin cans, metal drums, disused water storage pots and old tyres can be removed: this approach is often referred to as *source reduction*. Mosquito breeding in water storage pots that are in use can be reduced by covering up their openings, but this simple practice is not popular with the owners and it soon becomes neglected. The introduction of a reliable piped water supply should help reduce people's dependence on water storage containers and thereby reduce breeding of mosquitoes such as *Aedes aegypti*. However, in many areas having piped water people continue to store water in containers as insurance against an unreliable water supply.

In the Indian subcontinent water tanks are commonly sited on rooftops and are important breeding places of the vector (*Anopheles stephensi*) of urban malaria. Fitting these with mosquito screening would prevent breeding, but such covers usually become torn or are removed.

Some mosquitoes, such as *Culex quinquefasciatus*, breed in damaged septic tanks and soak-away pits, but this is prevented if the tanks and pits are repaired so that egg-laying females cannot gain access. This mosquito also commonly breeds in pit latrines, but this can be stopped if small (2–3 mm) expanded ***polystyrene beads*** are tipped into latrines to form a floating layer about 1–2 cm thick. This suffocates larvae and pupae and also prevents mosquitoes laying their eggs on the water surface.

Larval breeding places such as ponds, borrow pits, freshwater and saltwater marshes can be drained. An important advantage of filling in, draining or removing larval habitats is that such measures can lead to permanent control, but this approach is not always feasible. It is impossible, for example, to fill in all the scattered, small and temporary collections of water such as pools, vehicle tracks and puddles which may appear during the rainy season. Larger and more permanent habitats such as swamps may prove too costly to drain. Moreover, the local people may, understandably, not want certain breeding places filled in if the water is needed for domestic purposes or the sites used as watering points for livestock; such habitats may be an essential part of their life. The feasibility of eradicating breeding places must be assessed individually in each area.

Environmental manipulation   If it is not feasible to eliminate mosquito breeding places it may be possible to alter them to make them unsuitable as larval habitats. For example, some mosquitoes breed in isolated pools and small marshy areas that form at the edges of ditches and streams with winding courses. Realigning these water courses to increase water flow and prevent the build-up of static pockets of water can greatly reduce mosquito breeding. The periodic opening of sluice gates has been practised in India and Malaysia to flush out larvae from small isolated pools of water. Other environmental modifications include the removal of overhanging vegetation to reduce breeding by shade-loving mosquitoes; conversely planting vegetation along reservoirs and streams may eliminate sun-loving species. Intermittent flooding of ricefields to allow drying out every 3–5 days can substantially reduce populations of several important vectors. Removal of rooted or floating vegetation will prevent breeding of *Mansonia* species because they require plants to obtain their oxygen requirements.

Instead of draining marshy areas they can be excavated to form areas of relatively deep permanent water with well-defined vertical banks. This process is called ***impoundment***. It completely alters the habitat, making it unsuitable for many mosquitoes, especially *Aedes*, *Ochlerotatus* and *Psorophora* species, which lay their eggs on wet muddy edges of pools that are scattered over extensive marshy areas. Both small and large, freshwater and saltwater marshy areas can be converted into impounded waters. Sometimes such impounded waters are stocked with fish and ducks, which may also help reduce mosquito breeding.

There is the danger, however, that whenever a larval habitat has been modified to reduce breeding of certain mosquitoes the new conditions created may now support breeding of other mosquito species that were previously either absent or uncommon.

Chemical control

Most control directed against mosquitoes, except malaria vectors, consists of applying larvicides.

Oils   One of the oldest control methods consists of spraying mineral oils onto the water surface of breeding places to kill mosquito larvae by poisoning and suffocation. Diesel oil and kerosene (paraffin) applied at the high rate of 140–190 litres/ha have been used for over 100 years to control mosquito breeding, but because of environmental concerns smaller dosages of lighter mineral oils are recommended. The addition of detergents or 1–2.5% vegetable oils (e.g. castor oil, coconut oil) increases the spreading power of oils allowing application rates to be reduced to 18–50 litres/ha. Although such oils are still used, mainly in tropical countries, developed countries increasingly use specially formulated commercial high-spreading oils which are environmentally more friendly and are used at reduced dosage rates, commonly 9–27 litres/ha or less. Other useful larvicides include monomolecular films of lecithins and various non-petroleum oils which interfere with the properties of the air–water interface and cause larvae, pupae and even emerging or ovipositing adults to drown.

Oils have to be sprayed on breeding places about every 7–10 days in most tropical countries to ensure that larvae hatching from eggs are killed before they pupate and give rise to adults. Less frequent applications can be made in cooler temperate areas because the speed of development of the immature stages is much slower.

Paris Green   Another old control method is the application of a fine dust of Paris Green (copper acetoarsenite) to breeding places. When ingested by surface-feeding mosquito larvae, such as *Anopheles* species, it acts as a stomach poison. Paris Green can also be formulated as pellets or granules that either float on the water surface or sink. The latter formulation can be effective against culicine species whose larvae are mainly bottom-feeders. Although mosquito larvae have not developed resistance to Paris Green this insecticide has largely been replaced by the organophosphate and carbamate insecticides.

As with oils, repeated applications are usually needed.

Insecticides   With the arrival of residual insecticides such as DDT in the mid 1940s the use of oils and Paris Green was more or less abandoned in favour of spraying larval habitats with these more modern chemicals. However, because of their persistence in the environment and accumulation in food chains DDT and other organochlorine insecticides should not be used as larvicides. In their place, less persistent and biodegradable insecticides should be used, such as the organophosphates and carbamates.

Recommended chemicals for larviciding include organophosphates (e.g. malathion, pirimiphos methyl, fenitrothion and temephos), and carbamates (e.g. propoxur). Pyrethroids such as permethrin and deltamethrin can also be used as larvicides but because they tend to kill greater numbers of other aquatic insects, crustaceans and even fish they should be used with caution and only in special cases. In organically polluted waters insecticides are less effective and so either higher dosage rates must be used or the more effective organophosphates such as fenthion or chlorpyrifos applied. Chlorpyrifos is more toxic to mosquito larvae than most other insecticides, but it causes higher mortalities of fish and other aquatic organisms, and is consequently not so widely used.

All the above insecticides usually have to be *sprayed* on breeding places in tropical areas every 10–14 days, and more frequently on highly polluted waters.

Temephos has very low mammalian toxicity and briquettes, granules or microencapsulated formulations which slowly release the insecticide over days or even weeks can be placed in containers holding potable water to control *Aedes aegypti*. However, in Yangon, Myanmar, people have refused to have their water pots treated with temephos, because they consider any insecticide in drinking water as environmental contamination. This attitude is likely to spread.

Mansonia larvae can be killed by spraying herbicides to destroy the aquatic vegetation on which they rely to obtain their oxygen.

Larvicides are usually applied as emulsions or oil solutions, but granules (0.25–0.6 mm) or pellets (0.6–2 mm) are better for penetrating dense growths of aquatic vegetation. Insecticides formulated as *slow-release* granules or pellets can be scattered over marshy areas when they are relatively dry, then when they become flooded larvae hatching from drought-resistant aedine eggs, such as those of *Aedes*, *Ochlerotatus* and *Psorophora* species, are killed as the granules release their toxicants into the water. Larvicides are usually delivered from knapsack-type sprayers carried on the backs of operators, but they are sometimes dispersed from vehicle-mounted spraying machines. Large or inaccessible areas may require aerial spraying from helicopters or small fixed-wing aircraft.

Insect growth regulators (IGRs)

These are compounds, such as methoprene and pyriproxyfen, that arrest larval development of insects, or compounds, like diflubenzuron, which inhibit chitin formation in the immature stages. When used as larvicides these chemicals have the benefit of being environmentally friendly because they are more or less specific in killing mosquitoes and possess extremely low toxicity to humans. Methoprene is considered by the WHO to be safe for use in drinking water. It is the most commonly used IGR for mosquito control, and can be formulated as a liquid, granules or briquettes which can give control for more than 100 days. However, the relatively high cost of IGRs limits their use in poorer countries.

Integrated control

It has become fashionable to advocate *integrated* control, which usually means combining biological and insecticidal methods: for example, the introduction of predacious fish to breeding places which are also sprayed with insecticides that have minimum effect on the fish. However, it is better to regard integrated control as any approach that takes into consideration more than one method, whether these are directed at only the larvae or adults, or both.

1.5.2 Control directed at adults
Personal protection

Much can be done to reduce being bitten by mosquitoes. Houses, hospitals and other buildings can have windows and doors covered with mosquito screening, made of either strong plastic or non-corrosive metal. It is essential that screening is kept in good repair. Screens of 6–8 mesh (i.e. 6–8 holes/cm) will exclude most mosquitoes. Finer-mesh screening will keep out smaller biting flies, some of which may be vectors, but will appreciably reduce ventilation and light. If houses are unscreened, or if screening is defective, *mosquito nets* of 9–10 mesh can be used to protect against night-biting mosquitoes. The actual size of the holes depends on the thickness of the thread, as well as their number; generally their size is 1.2–1.5 mm. Nets should be tucked in under mattresses or bedding, never allowed to drape loosely over beds. Torn nets are useless unless they have been impregnated with pyrethroid insecticides such as permethrin (see p. 48). Nets should be placed over beds before sunset. The main disadvantage of nets is that they reduce ventilation.

Small spray-guns (e.g. flit-guns) filled with pyrethrum or permethrin dissolved in kerosene, or alternatively pressurized aerosol canisters containing pyrethroids, are commonly used to spray bedrooms early in the evenings. Mosquito coils impregnated with pyrethroid insecticides, especially fast-acting ones like bioallethrin, which when ignited smoulder for

6–10 hours to produce an insecticidal smoke, are commonly used in tropical countries. A more sophisticated method is to place small insecticide-impregnated tablets (called vaporizing or fumigant mats) on a mains-operated electric mini-heater.

Suitable insect repellents are DEET, DIMP and a newer compound 'Autan' that is based on piperidines and is as effective as DEET but (unlike DEET) does not attack plastics. Under optimal conditions these repellents can provide protection for 6–10 hours. Citronella oil can give good protection against mosquitoes, but only for about an hour. Sweating and rubbing usually reduce the period of effectiveness. Repellents are applied to the hands, arms, neck and face (taking care to avoid the eyes), and the ankles and legs, irrespective of whether socks or long trousers are worn. Locally made repellent soap (incorporating 20% DEET and 0.5% permethrin) has been shown to be effective, although the greasy feeling produced is not always acceptable. Repellent- or insecticide-impregnated (e.g. permethrin) clothing, such as wide-mesh jackets and hoods (as used by military personnel) give longer protection than repellents applied to the skin. If treated clothing is kept in plastic bags when not in use it should remain effective for many months before re-impregnation is needed.

Aerosols, mists and fogs

Motorized knapsack mist-blowers, shoulder-carried thermal foggers or vehicle-mounted machines that generate insecticidal aerosols (<50 μm) or mists ($51–100$ μm) can be used to kill outdoor resting (exophilic) adult mosquitoes. Fogs are produced when very fine aerosol droplets ($5–15$ μm) are so numerous that they substantially reduce visibility. Indoor resting (endophilic) adults are also occasionally killed by mist-blowers or thermal foggers. Several insecticides can be used, including malathion, fenitrothion, pirimiphos methyl, bendiocarb and the synthetic pyrethroids. Although such applications can be spectacular there is very little residual effect and areas cleared of adult mosquitoes are rapidly invaded by newly emerged adults and mosquitoes flying in from outside the treated area. Repeated applications are needed to sustain control.

Applications of aerosols and mists are best made in calm weather, and usually in the evenings or early mornings when there are fewer thermals rising from the ground and less turbulence. Aerial applications from helicopters or fixed-wing aircraft usually give better coverage and more effective control than ground-based operations.

Ultra-low-volume applications   Ultra-low-volume (ULV) techniques

apply the minimum (<5 litres/ha) of concentrated insecticides, often just 225–500 ml/ha, as against 5–25 litres/ha with conventional spraying. This means that trucks or aircraft can spray much larger areas on a tank of insecticides before they have to return to base. Insecticides commonly used in ULV operations include malathion, pirimiphos methyl, fenitrothion, propoxur, chlorpyrifos and the pyrethroids. With aerial applications *droplet size* of the insecticide is bigger (150–200 μm) than that used in ground-based applications (50–100 μm) because they decrease in size, due to evaporation, as they fall to the ground. Generally the size of droplets hitting mosquitoes should be 15–25 μm. In addition to rapidly reducing outdoor resting and biting mosquito populations ULV spraying is used in potential or actual epidemic situations to prevent or control disease outbreaks. In emergency situations aerial spraying gives the fastest and most effective control of vectors, and has been used to stop transmission of haemorrhagic dengue, Japanese encephalitis and in North America encephalitis viruses.

Residual house-spraying

Some mosquitoes, such as many malaria vectors and *Culex quinquefasciatus*, rest in houses before and/or after blood-feeding. Their populations can be reduced by insecticidal spraying of houses, but as this approach is mainly used in malaria control operations it is described in Chapter 2.

FURTHER READING

Anon. (1995) Vector control without chemicals: has it a future? A symposium. *Journal of the American Mosquito Control Association*, **11**, 247–93.

Bates, M. (1949) *The Natural History of Mosquitoes*. New York, NY: Macmillan. (Facsimile. New York, NY: Harper and Row, 1965.)

Bock, G. R. and Cardew, G. (eds.) (1996) *Olfaction in Mosquito-Host Interactions*. Chichester: Wiley.

Bowen, M. F. (1991) The sensory physiology of host-seeking behavior in mosquitoes. *Annual Review of Entomology*, **36**, 139–58.

Chapman, H. C. (ed.) (1985) Biological control of mosquitoes. *American Mosquito Control Association Bulletin*, **6**, 1–218.

Clark, G. G. (coordinator) (1994) Prevention of tropical diseases: status of new and emerging vector control strategies. Proceedings of a symposium on vector control. *American Journal of Tropical Medicine and Hygiene*, **50** (6) (suppl.), 1–159.

Clements, A. N. (1992) *The Biology of Mosquitoes. Volume 1: Development, Nutrition and Reproduction*. London: Chapman and Hall.
(1999) *The Biology of Mosquitoes. Volume 2: Sensory Reception and Behaviour*. Wallingford: CABI.

Curtis, C. F. (ed.) (1989) *Appropriate Technology in Vector Control*. Boca Raton, FL: CRC Press.

de Barjac, H. and Sutherland, D. J. (eds.) (1990) *Bacterial Control of Mosquitoes and Black Flies: Biochemistry, Genetics and Applications of* Bacillus thuringiensis *and* Bacillus sphaericus. New Brunswick, NJ: Rutgers University Press.

Foster, W. A. and Walker, E. D. (2002) Mosquitoes (*Culicidae*). In: *Medical and Veterinary Entomology*, ed. G. Mullen, G. and L. Durden. Amsterdam: Academic Press, pp. 203–62.

Gillett, J. D. (1971) *Mosquitos*. London: Weidenfeld and Nicolson.

Harbach, R. and Sandlant, G. (1997) *CABIKEY: Mosquito Genera of the World*. CD-ROM for Windows 3x or Windows 95. Wallingford: CABI.

Horsfall, W. R. (1972) *Mosquitoes: their Bionomics and Relation to Disease*. New York, NY: Hafner.

Lacey, L. A. and Lacey, C. M. (1990) The medical importance of riceland mosquitoes and their control using alternatives to chemical insecticides. *Journal of the American Mosquito Control Association*, **6** (suppl. 2), 1–93.

Laird, M. (1988) *The Natural History of Larval Mosquito Habitats*. London: Academic Press.

Laird, M. and Miles, J. W. (eds.) (1983) *Integrated Mosquito Control Methodologies. Volume 1: Experience and Components from Conventional Chemical Control*. London: Academic Press.

(eds.) (1985) *Integrated Mosquito Control Methodologies. Volume 2: Biocontrol and Other Innovative Components, and Future Directions*. London: Academic Press.

Ribeiro, J. M. C. (1987) Role of saliva in blood-feeding by arthropods. *Annual Review of Entomology*, **32**, 463–78.

Schaffner, E., Angel, G., Geoffroy, B., Hervy, J.-P., Rhaiem, A. and Brunhes, J. (2000) *The Mosquitoes of Europe: an Identification and Training Programme*. CD-ROM. Windows 9x, NT, ME, 2000. Montpellier: EID. (In French and English.)

Service, M. W. (1989) Rice, a challenge to health. *Parasitology Today*, **5**, 162–5.

(1993) Mosquitoes (Culicidae). In: *Medical Insects and Arachnids*, ed. R. P. Lane and R. W. Crosskey. London: Chapman and Hall, pp. 120–40.

(1993) *Mosquito Ecology: Field Sampling Methods*. 2nd edn. London: Chapman and Hall.

Spielman, A. and d'Antonio, M. (2001) *Mosquito: a Natural History of Our Most Persistent and Deadly Foe*. London: Faber and Faber.

World Health Organization (1982) Manual on environmental management for mosquito control with special emphasis on malaria vectors. *WHO Offset Publication*, **66**, 1–238.

(1984) *Chemical Methods for the Control of Arthropod Vectors and Pests of Public Health Importance*. Geneva: World Health Organization.

(1990) *Equipment for Vector Control*. 3rd edn. Geneva: World Health Organization.

(1992) Vector resistance to pesticides. Fifteenth report of the WHO Expert Committee on Vector Biology and Control. *World Health Organization Technical Report Series*, **818**, 1–71.

(1996) *Operational Manual on the Application of Insecticides for Control of the Mosquito Vectors of Malaria and Other Diseases*. WHO/CTD/VBC/96.1000. Geneva: World Health Organization.

(1997) *Vector Control: Methods for Use by Individuals and Communities*, prepared by J. A. Rozendaal. Geneva: World Health Organization.

See also references at the ends of Chapters 2 and 3.

2

Anopheline mosquitoes (Anophelinae)

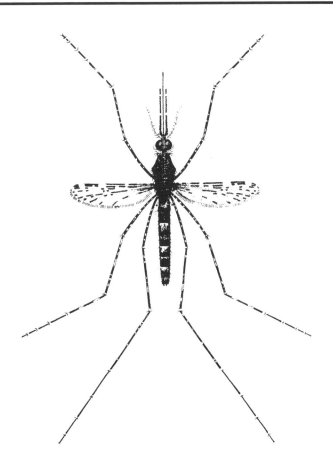

The subfamily Anophelinae contains three genera, but as explained in Chapter 1 only the genus *Anopheles* is of medical importance. *Anopheles* mosquitoes have a worldwide distribution, occurring not only in tropical areas but also in temperate regions. There are about 430 different species. The most important disease carried by *Anopheles* mosquitoes is malaria. Some *Anopheles* species are also vectors of filariasis, especially that caused by *Wuchereria bancrofti*, but some also transmit *Brugia malayi* and *Brugia timori*. A few species transmit arboviruses that are of minor medical importance.

2.1   EXTERNAL MORPHOLOGY OF *ANOPHELES*

The main features distinguishing adults of the Anophelinae, and in particular the genus *Anopheles*, from other mosquitoes have been given in Chapter 1, but a brief summary is presented here.

Eggs are laid singly and have air-filled *floats* (Fig. 1.8) that help them float on the water surface.

Larvae do not have a siphon and consequently lie *parallel* to the water surface (Fig. 1.13). Dorsally a *tergal plate* and paired *palmate hairs* are present on most abdominal segments (Fig. 1.9).

In the pupae the abdominal segments have numerous short setae, and segments 2–7 or 3–7 have in addition short peg-like spines (Fig. 1.11) which are absent in culicines.

Most, but not all, *Anopheles* have *spotted wings*, that is the dark and pale scales are arranged in small blocks or areas on the veins (Fig. 1.12). The number, length and arrangement of these dark and pale areas differ considerably in different species and provide useful characters for species identification. Unlike Culicinae the dorsal and ventral surfaces of the abdomen are almost, or entirely, devoid of appressed scales. In both sexes the palps are about as long as the proboscis and in males, but not females, they are enlarged (that is clubbed) apically (Fig. 1.13). See page 18 for minor differences distinguishing anophelines from culicines.

2.2   LIFE CYCLE OF *ANOPHELES*

After mating and blood-feeding *Anopheles* lay some 50–200 small brown or blackish boat-shaped eggs (Fig. 1.8) on the water surface. *Anopheles* eggs cannot withstand desiccation and in tropical countries hatch within 2–3 days, but in colder temperate climates hatching may not occur until after about 2–3 weeks, the duration depending on temperature.

As in all mosquitoes there are four larval instars. *Anopheles* larvae are filter-feeders and unless disturbed remain at the water surface, feeding on bacteria, yeasts, protozoans and other micro-organisms, and also breathing in air through their spiracles. When feeding, larvae rotate their heads through 180° so that the ventrally positioned mouthbrushes can sweep the underside of the water surface. Larvae are easily disturbed by shadows or vibrations and respond by swimming quickly to the bottom of the water. They resurface some seconds or minutes afterwards.

Anopheles larvae occur in many different types of large and more or less permanent habitats, ranging from freshwater and saltwater marshes, mangrove swamps, grassy ditches, ricefields, edges of streams and rivers to ponds and borrow pits. They are also found in small and often temporary breeding places such as puddles, hoofprints, wells, discarded tin cans and sometimes water storage pots. A few species occur in water-filled tree-holes. In the Neotropical region (Central and South America and the West Indies) a few *Anopheles* breed in water that collects in the leaf axils of epiphytic plants such as **bromeliads** growing on tree branches. Bromeliads somewhat resemble pineapple plants. Some *Anopheles* prefer habitats with aquatic vegetation while others favour habitats without vegetation; some species like exposed sunlit waters whereas others prefer more shaded breeding places. In general *Anopheles* prefer clean and unpolluted waters and are usually absent from habitats that contain rotting plants or are contaminated with faeces.

In tropical countries the larval period frequently lasts only about 7 days, but in cooler climates the larval period may be 2–4 weeks. In temperate areas some *Anopheles* overwinter as larvae and consequently may live many months.

The comma-shaped pupae normally remain floating at the water surface with the aid of the pair of palmate hairs on the cephalothorax, but when disturbed they swim vigorously down to the bottom with characteristic jerky movements. The pupal period lasts 2–3 days in tropical countries but sometimes as long as 1–2 weeks in cooler climates, after which the adult mosquito emerges.

2.2.1   Adult biology and behaviour

Most *Anopheles* are **crepuscular** or **nocturnal** in their activities. Thus blood-feeding and oviposition normally occur in the evenings, at night or in the early mornings around sunrise. Some species such as *An. albimanus*, a malaria vector in Central and South America, bite people mainly outdoors (**exophagic**) from about sunset to 21:00 hours. In contrast, in Africa species of the *An. gambiae* complex, which contains probably the world's most efficient malaria vectors, bite mainly after 23:00 hours and mostly indoors (**endophagic**). As already discussed in Chapter 1, the times of biting, and whether adult mosquitoes are exophagic or endophagic, may be important in the epidemiology of diseases.

Both before and after blood-feeding some species rest in houses (**endophilic**), whereas others rest outdoors (**exophilic**) in a variety of shelters, such as amongst vegetation, in rodent burrows, in cracks and crevices in trees, under bridges, in termite mounds, in caves and rock fissures, and in cracks in the ground. Most *Anopheles* species are not exclusively exophagic or endophagic, exophilic or endophilic, but exhibit a mixture of these extremes of behaviour. Similarly, few *Anopheles* feed exclusively on either humans or non-humans, most feeding on both people and animals, but the degree of anthropophagism and zoophagism varies according to species. For example, *An. culicifacies*, an important Indian malaria vector, commonly feeds on cattle as well as humans, whereas in Africa *An. gambiae* sensu stricto (a species of the *gambiae* complex) feeds more rarely on cattle and thus maintains a stronger mosquito–man contact. This is one of the reasons why *An. gambiae* is a more efficient malaria vector than *An. culicifacies*.

2.3   MEDICAL IMPORTANCE

2.3.1   Biting nuisance

Although *Anopheles* mosquitoes may not be disease vectors in an area they may nevertheless constitute a biting nuisance. Usually, however, it is the culicine mosquitoes, especially *Aedes, Ochlerotatus* and *Psorophora* species, that cause biting problems.

2.3.2   Malaria

Only mosquitoes of the genus *Anopheles* transmit the four parasites (*Plasmodium falciparum, P. vivax, P. malariae, P. ovale*) that cause human malaria. Because the sexual cycle of the malaria parasite occurs in the vector, it is conventional to call the mosquito the **definitive host**, and humans the intermediate host.

Male and female malaria gametocytes are ingested by female mosquitoes during blood-feeding and pass to the mosquito's stomach where they undergo cyclical development that includes a sexual cycle termed sporogony. Only **gametocytes** survive in the mosquito's stomach, all other blood forms of the malaria parasites (the asexual forms) are destroyed. Male gametocytes (microgametocytes) extrude flagella which are the male gametes (microgametes), and the process is called **exflagellation**. The microgametes break free and fertilize the female gametes (macrogametes) which have formed from the macrogametocytes. As a result of fertilization a **zygote** is formed, which elongates to become an ookinete. This penetrates the wall of the mosquito's stomach and reaches its outer membrane where it becomes spherical and develops into an **oocyst**, which can be seen on the stomach walls of vectors about 4–5 days after an infective blood-meal. The nucleus of the oocyst divides repeatedly to produce numerous spindle-shaped **sporozoites**. After about 8 days when the oocyst is fully grown (about 40–80 μm) it ruptures and thousands of sporozoites are released into the haemocoel of the mosquito. The sporozoites (10–15 μm) are carried in the insect's haemolymph to all parts of the body but most penetrate the **salivary glands** where they are usually found after 9–14 days. However, the time required for this cyclical development (*extrinsic cycle*), depends on both temperature and *Plasmodium* species. At 30 °C sporogony in *P. falciparum* takes 9 days, at 25 °C 10 days, at 25 °C 11 days and at 20 °C 23 days, while below 18 °C it cannot be completed. With *Plasmodium vivax* sporogony develops faster: it is completed in 9 days at 25 °C and in 16 days at 20 °C.

The mosquito is now infective and sporozoites are inoculated into people the next time the mosquito bites. A single oocyst produces 1000 or more sporozoites, and it has been estimated that in heavy infections there may be as many as 60 000–70 000 sporozoites in the vector's salivary glands. However, the number may be much smaller than this, and very few (sometimes just 5–10) are actually injected into a person during feeding.

The *sporozoite rate*, that is the percentage of female vectors with sporozoites in the salivary glands, varies considerably, not only from species to species of mosquito but also according to locality and season. Sporozoite rates are often about 1–5% in species such as *An. gambiae* and *An. arabiensis* of the *An. gambiae* complex, but less than 1% in many other species such as *An. albimanus* and *An. culicifacies*. For practical purposes it can be said that once a vector becomes infective it remains so throughout its life.

2.3.3 Important malaria vectors

Although there are some 430 species of *Anopheles* only about 70 are malaria vectors and of these probably only about 40 are important. Malaria vectors are often divided into primary and secondary vectors, but this can be misleading because a species may be classed as a primary vector in some areas but be only a secondary vector in others. Although presenting a list of important malaria vectors must be somewhat subjective I have nevertheless attempted to do this and given notes on their principal larval habitats and biting behaviour. These notes are no more than a guide to their behaviour, which may vary in different parts of their geographical distribution. Several species are found in species complexes, which comprise virtually identical-looking species that can be separated only by their chromosomal banding patterns, by biochemical procedures or by molecular methods. However, species within a complex may differ in their biology, behaviour, distribution and vector status. Some of the more important species complex are referred to in this list of vectors. The most important and best known is the *An. gambiae* complex in Africa.

(1) Sub-Saharan Africa

Anopheles gambiae   This is the most anthropophagic species of the seven that are in the *An. gambiae* complex and the most important malaria vector – it is probably the world's most efficient vector. Larval habitats are sunlit pools, puddles, borrow pits and ricefields. Aduts bite humans both indoors and outdoors, and also feed on domestic animals, resting mainly indoors, but also outdoors. Other malaria vectors in the *An. gambiae* complex are *An. arabiensis, An. melas, An. merus* and *An. bwambae* (see below). *An. quadriannulatus* and *An. quadriannulatus* species B, also in the complex, bite cattle not humans so are not vectors.

Anopheles arabiensis This is another important malaria vector. Larvae in same types of habitat as *An. gambiae,* namely ricefields, borrow pits and also temporary waters such as pools, puddles and hoofprints. Adults bite humans indoors and outdoors but also cattle; after feeding they rest either indoors or outdoors. This species tends to occur in drier areas than does *An. gambiae,* and is more likely to bite cattle and rest outdoors than *An. gambiae.*

Anopheles melas and *Anopheles merus* *Anopheles melas* breeds in coastal salt waters such as mangrove swamps; it is restricted to West Africa. *Anopheles merus* is a coastal saltwater species in East and southern Africa, but can be found breeding in inland saltwater habitats. Adults of both species bite humans and rest indoors and outdoors; they are both regarded as secondary malaria vectors.

Anopheles bwambae Restricted to breeding in warm mineral springs in Uganda; it is a rare species and not considered an important vector although locally it can transmit malaria.

Anopheles funestus Larvae occur in more-or-less permanent waters, especially those with vegetation, such as marshes, edges of streams, rivers and ditches, and ricefields with mature plants providing shade. Prefers shaded habitats. Adults bite humans predominantly, but also domestic animals; feeds indoors and also outdoors; after feeding adults rest mainly indoors.

(2) Europe, North Africa and the Middle East
Anopheles atroparvus One of nine species in the *An. maculipennis* complex. Sunlit and exposed pools and ditches with either fresh or brackish water, also ricefields. Adults bite humans and domestic animals and usually rest in stables, cow sheds and piggeries. Adults hibernate in these and other shelters during the winter, but periodically emerge to take blood-meals.

Anopheles labranchiae One of nine species in the *An. maculipennis* complex. Brackish waters of coastal marshes, freshwater marshes, edges of grassy streams and ditches, ricefields; prefers sunlight. Bites humans and domestic animals indoors and outdoors; rests mainly in houses or animal shelters after feeding. Overwinters as hibernating adults.

Anopheles pharoensis Marshes, ponds, especially those with abundant grassy or floating vegetation, also ricefields. Adults bite humans and animals indoors or outdoors, and rest outdoors after feeding. Can be an important vector in Egypt

Anopheles sacharovi Fresh or brackish waters of coastal or inland marshes, pools and ponds, especially those with vegetation. Prefers sunlit habitats. Bites humans and animals indoors or outdoors; usually rests in houses or animal shelters after feeding.

Anopheles sergentii Borrow pits, ricefields, ditches, seepage waters, slow-flowing streams, sunlit or partially shaded habitats. Adults bite humans or animals indoors or outdoors, resting in houses and caves after feeding.

Anopheles stephensi Possibly a species within a complex. *Anopheles stephensi* can be an important vector locally, especially in towns. Apart from being found in Egypt, Iraq, Iran and Saudi Arabia it is common in the Indian subcontinent where it is commonly the main vector of urban malaria. Larvae breed in fresh, brackish or even polluted waters in man-made habitats such as water tanks, cisterns, wells, gutters, water storage jars and containers. Adults bite humans indoors or outdoors, and rest mainly indoors afterwards. *Anopheles stephensi* has a very wide distribution extending from the Middle East across Pakistan and India, to Myanmar, Thailand and China.

Anopheles superpictus Flowing waters such as torrents of shallow water over rocky streams, pools in rivers, muddy hill streams. Vegetation may be present; prefers sunlight. Bites humans and animals indoors and outdoors, and after feeding rests mainly in houses and animal shelters, but also in caves.

(3) Indian subcontinent
Anopheles culicifacies One of five species in the *An. culicifacies* complex. Larvae occur in a variety of clean and unpolluted habitats, irrigation ditches, pools, wells, borrow pits, edges of streams, marshes, ricefields and occasionally in brackish waters. Prefers domestic animals but commonly bites humans indoors or outdoors; rests mainly indoors after feeding. Is the main malaria vector in much of the region.

Anopheles fluviatilis One of four species in the *An. fluviatilis* complex. Flowing waters such as hill streams, pools in river beds, irrigation ditches, seepages; prefers sunlight. Bites humans and domestic animals; feeds and rests either indoors or outdoors.

Anopheles minimus   One of four species in the *An. minimus* complex. Flowing waters such as foothill streams, springs, irrigation ditches, seepages, borrow pits and ricefields; prefers shaded areas. Feeds mainly on humans, but will bite domestic animals; feeds and rests mainly indoors.

Anopheles stephensi   See Europe, North Africa and the Middle East, above.

Anopheles sundaicus   One of three species in the *An. sundaicus* complex. Salt or brackish waters including lagoons, marshes, pools and seepages, especially with putrefying algae and aquatic weeds. Mainly a coastal species but found in freshwater inland pools in Java and Sumatra. Prefers sunlit habitats. Bites humans and domestic animals indoors and outdoors; rests mainly indoors after feeding.

Anopheles superpictus   See Europe, North Africa and the Middle East, above.

(4) South-east Asia

Anopheles aconitus   Ricefields, swamps, irrigation ditches, pools and streams with vegetation; prefers sunlit habitats. Adults feed indoors or outdoors on humans but also commonly on animals; adults rest indoors or outdoors after feeding.

Anopheles anthropophagus   Shaded pools and ponds, only rarely in ricefields. Adults bite humans indoors, and rest mainly indoors after feeding.

Anopheles balabacensis   Muddy and shaded forest pools, animal hoofprints and vehicle ruts, occasionally in deep wells. Adults bite humans and cattle; feeds and rests outdoors. Morphologically and biologically very similar to *An. dirus*, but has a more restricted distribution – Sabah, Java, Borneo and certain Philippine islands.

Anopheles campestris   Deep and usually shaded or partially shaded waters such as ditches, wells, and shaded parts of ricefields; larvae also sometimes in brackish waters. Bites humans and animals indoors and outdoors; substantial numbers rest indoors after feeding.

Anopheles culicifacies   See Indian subcontinent, above.

Anopheles dirus   One of seven species in the *An. dirus* complex. Shaded pools, hoofprints in or at the edges of forests. Adults bite humans and domestic animals, mainly outdoors, and stay outdoors after feeding.

Morphologically and biologically very similar to *An. balabacensis*, but has a more widespread distribution from western India and Bangladesh to South-east Asia.

Anopheles donaldi   Shaded habitats such as tree-covered swamps, forest pools, often with vegetation, ricefields. Bites domestic livestock but also feeds on humans inside or outside houses; rests mainly outdoors.

Anopheles flavirostris   Flowing waters such as foothill streams, springs, irrigation ditches, also borrow pits and ricefields. Prefers shaded areas of sunlit habitats. Feeds mainly on humans, but also on domestic animals; feeds indoors or outdoors but rests mainly outdoors after feeding.

Anopheles fluviatilis   See Indian subcontinent, above.

Anopheles letifer   Often in acidic stagnant waters such as pools, swamps and ponds, especially on coastal plains; prefers shade. Adults bite livestock and humans mainly outdoors, and rest afterwards outdoors.

Anopheles leucosphyrus   One of two species in the *An. leucosphyrus* complex. Clear seepage pools in forests. Adults bite humans inside and outside houses, but afterwards rest outdoors. This species is morphologically and biologically similar to *An. balabacensis* and *An. dirus*.

Anopheles maculatus   One of eight species in the *An. maculatus* complex. Seepage waters, pools formed in streams, edges of ponds, ditches and swamps with much vegetation; prefers sunlight. Bites humans and animals mainly outdoors and rests outdoors after feeding.

Anopheles minimus   See Indian subcontinent, above.

Anopheles nigerrimus   Larvae in deep ponds, ricefields, irrigation ditches and marshes with much vegetation; prefers sunlight. Adults bite humans and animals mainly outdoors and rest mainly outdoors.

Anopheles sinensis   Has been confused with *An. anthropophagus*. Common in China where in the north it may be the only malaria vector. Larvae in ricefields, marshes, ditches and grassy ponds. Adults bite cattle and humans, indoors or outdoors, rests indoors or outdoors.

Anopheles subpictus   One of four species in the *An. subpictus* complex. Muddy pools near houses, borrow pits, gutters, also brackish waters. Bites mainly animals, but also humans both indoors and outdoors; rests indoors or outdoors after feeding.

Anopheles sundaicus   See Indian subcontinent, above.

(5) Mexico and Central America

Anopheles albimanus   Fresh or brackish waters such as pools, puddles, marshes, ponds and lagoons, especially those containing floating or grassy vegetation; prefers sunlit habitats. Adults feed on humans and domestic animals both indoors and outdoors; after feeding adults rest mainly indoors.

Anopheles albitarsis   One of four species in the *An. albitarsis* complex. Larvae nearly always in sunlit ponds, large pools and marshes with filamentous algae. Bites humans and domestic animals almost indiscriminately. Feeds outdoors and also indoors, and usually rests outdoors after feeding.

Anopheles aquasalis   Tidal saltwater marshes, lagoons, saltwater regions of rivers, estuaries, rarely in fresh water; sunlit or shaded habitats. Adults bite humans and domestic animals indoors or outdoors, and rest mainly outdoors.

Anopheles darlingi   Freshwater marshes, lagoons, ricefields, swamps, lakes, ponds, pools, edges of streams, especially with vegetation. Mainly shaded larval habitats. Feeds mainly on humans indoors; stays indoors after feeding.

Anopheles pseudopunctipennis   Pools, puddles, seepage waters and edges of streams, especially habitats with algae; prefers sunlight. Feeds almost indiscriminately indoors or outdoors on humans and domestic animals; rests outdoors.

Anopheles punctimacula   Small pools, swamps, grassy pools at edges of streams; prefers shade. Bites humans and domestic animals both indoors and outdoors; rests indoors or outdoors after feeding.

(6) South America

Anopheles albitarsis   See Mexico and Central America, above.

Anopheles aquasalis   See Mexico and Central America, above.

Anopheles bellator   Larvae are found only in water collected in the leaf axils of bromeliads, which are epiphytes on trees in the Americas. Adults bite humans during the daytime in shaded forests, and also at night and may enter houses to feed. Adults rest mainly outdoors. The species will also bite domestic animals. It occurs in Trinidad, Venezuela, Surinam, Guyana and Brazil, where it can be a local malaria vector.

Anopheles cruzii   Another malaria vector that also breeds in bromeliad axils. Adults bite humans indoors and outdoors, and rest indoors and outdoors. Mainly a malaria vector in coastal areas of Brazil.

Anopheles darlingi   See Mexico and Central America, above.

Anopheles nuneztovari   Possibly one of three species in an *An. nuneztovari* complex. Muddy waters of pools, vehicle tracks, hoofprints and small ponds, especially in and around towns; prefers sunlight. Feeds mainly on animals but in northern Colombia and western Venezuela adults bite humans indoors and outdoors, and rest outdoors after feeding.

Anopheles punctimacula   See Mexico and Central America, above.

(7) Australasia
Anopheles farauti   *Anopheles farauti* is one of seven very closely related species forming the *farauti* group, which together form part of the 11 species in the *An. punctulatus* complex. Larvae usually occur in semi-permanent waters such as swamps, ponds, lagoons and edges of slow-flowing streams, but also in puddles, hoofprints, pools and man-made containers; water may be fresh, slightly brackish or even polluted, and in sunlight or shade. Adults bite animals but also humans indoors or outdoors, and rest mainly outdoors, but also indoors.

Anopheles koliensis   A species in the *An. punctulatus* complex. Larvae occur in marshy pools, irrigation ditches, pools at edges of forest streams, often in sunlit habitats. Adults bite humans and more occasionally animals, and after feeding rest mainly indoors, but they rest outdoors in some areas.

Anopheles punctulatus   One of three very similar species in the *An. punctulatus* complex. Larvae occur in temporary and often muddy pools, puddles, hoofprints and ditches. Adults bite humans in preference to animals, and rest indoors or outdoors after feeding.

2.3.4   Filariasis
Certain *Anopheles* species transmit filarial worms of *Wuchereria bancrofti*, *Brugia malayi* and *Brugia timori*, all of which cause filariasis in humans. The role of culicine mosquitoes as vectors of the first two species is described in Chapter 3, and details of principal mosquito vectors are given in Table 3.1.

Wuchereria bancrofti causes filariasis in people living in many tropical regions of the world (e.g. much of sub-Saharan Africa, Egypt, India and elsewhere in Asia including the Pacific area, Haiti, the Dominican Republic and small coastal foci in Brazil and Guyana). In many of these areas bancroftian filariasis is mainly an urban disease. In contrast *B. malayi* is more a rural disease and has a much more restricted distribution, occurring in Asian countries such as southern India, Malaysia, Vietnam, Indonesia, Thailand, Papua New Guinea, Philippines, Korea, and with small foci in China. It is absent from Africa and the Americas.

Both bancroftian and brugian filariasis occur in two basic forms: ***nocturnal periodic*** and ***nocturnal subperiodic***. Anophelines, together with certain culicines (*Culex quinquefasciatus*, various species of *Mansonia* and a few *Ochlerotatus* species) are the vectors of the nocturnal periodic form. Culicines are vectors of subperiodic forms (see Chapter 3).

In the nocturnal periodic form of these two parasites most of the microfilariae during the day are in the blood vessels supplying the lungs. At night, especially during the middle part, microfilariae migrate to the peripheral blood system and lymph vessels. Because of this marked 24-hour periodicity microfilariae are ingested mainly by night-biting mosquitoes such as *Anopheles*.

More than 25 anopheline species are known vectors of bancroftian filariasis, many of which are also the principal vectors of malaria. The anophelines involved in filariasis transmission differ according to the area. For example, vectors of nocturnally periodic *W. bancrofti* include *An. albimanus* (tropical Americas), *An. arabiensis*, *An. funestus*, *An. gambiae* (sub-Saharan Africa), *An. anthropophagus*, *An. balabacensis*, *An. dirus*, *An. flavirostris*, *An. letifer*, *An. leucosphyrus*, *An. maculatus*, *An. sinensis* and *An. subpictus* (India and South-east Asia), and *An. koliensis* and *An. punctulatus* (Papua New Guinea). There are no known animal reservoirs of the nocturnal periodic form of filariasis.

Nocturnal periodic *B. malayi* in Asia is transmitted by at least 10 anopheline species, including *An. anthropophagus*, *An. barbirostris*, *An. campestris*, *An. donaldi* and *An. sinensis*. There are no important animal reservoirs, although it is possible that some exist.

Brugia timori is known only from the small Indonesian islands of Timor and Flores, and from lowland areas of small neighbouring islands east of Java. Its microfilariae are nocturnally periodic and are transmitted by *An. barbirostris*, and possibly by other anopheline species. There are no known animal reservoirs.

Filarial development in mosquitoes

Development of all three filarial species in the mosquito vector, and the basic mode of transmission from mosquito to humans, are the same for all vectors. Basically, the life cycle in the mosquito is as follows. ***Microfilariae*** ingested with the blood-meal pass into the stomach of the mosquito (in some vectors such as *Anopheles* many may be destroyed during their passage through the oesophagus). Within 15–30 minutes they exsheath, penetrate the stomach wall and pass into the *haemocoel*, from where they migrate to the *thoracic muscles* of the mosquito. In the thorax the small larvae become more or less inactive, grow shorter but considerably fatter and develop, after 2 days, into 'sausage-shaped' forms. They undergo two moults and the resultant *third-stage* larvae become active, leave the muscles and migrate through the head and down the fleshy labium of the proboscis. This is the infective stage and is formed some 10 days or more after the microfilariae have been ingested with a blood-meal.

When the mosquito takes further blood-meals, several infective (third-stage) larvae (1.2–1.6 mm long) rupture the skin of the labella of the labium and crawl onto the surface of the host's skin, but many of them die. Only a few manage to find a skin abrasion, sometimes the small lesion caused by the mosquito's bite, and thus enter the skin and pass to the host's lymphatic system. It should be noted that the salivary glands are not involved in the transmission of filariasis, and also that there is no multiplication or sexual cycle of the parasites in the mosquito.

Infection rates of infective larvae in anopheline vectors vary according to the mosquito species and local conditions, but they are often about 0.1–5% for *W. bancrofti* and about 0.1–3% for *B. malayi*.

There are no animal reservoirs of *W. bancrofti*, but the nocturnal subperiodic form of *B. malayi*, transmitted by *Mansonia* mosquitoes, is a zoonosis (see Chapter 3).

The presence of filarial worms in the thoracic muscles of mosquitoes, or infective worms in the proboscis, does not necessarily implicate mosquitoes as vectors of bancroftian or brugian filariasis. This is because there are several other mosquito-transmitted filariae. For example, various *Setaria* species infect cattle, *Dirofilaria repens* and *Dirofilaria immitis* infect dogs and humans, and various other species of *Brugia*, such as *B. patei* in Africa and *B. pahangi* in Asia infect animals but not people. Careful examination is therefore essential to identify whether the filarial parasites found in mosquitoes as those of *W. bancrofti* or *B. malayi*.

Table 3.1 (p. 71) summarizes the distribution and vectors of filariasis.

2.3.5 Arboviruses

The word arbovirus is derived from the term 'arthropod-*borne virus*'. An arbovirus infection in either human or non-human hosts produces *viraemia* and the virus is ingested by blood-sucking insects such as mosquitoes when they take blood-meals. Within the vector the virus undergoes multiplication and/or cyclical development before being transmitted by the infected arthropod during refeeding. An arbovirus therefore undergoes obligatory development in an arthropod host. Yellow fever and dengue are typical arboviruses transmitted by *Aedes* mosquitoes. In contrast the virus causing poliomyelitis is not an arbovirus for although it can be transmitted by certain flies, such as house-flies, this is purely mechanical transmission; the virus does not undergo any multiplication and/or development in an arthropod. Time taken for an infected mosquito to become infective, that is virus is present in the salivary glands, is called the ***extrinsic incubation period***, and varies according to temperature and the species of both arbovirus and mosquito. Over 500 arboviruses have been catalogued, but little is known about the epidemiology of many of them, and not all may be real arboviruses. About 125 arboviruses infecting humans are transmitted by mosquitoes, mostly by culicines, in particular by *Aedes*, *Ochlerotatus* and *Culex* species. Other arboviruses are spread by ticks and other arthropods.

In 1959 a new painful, but non-fatal, arboviral disease called o'nyong nyong (a local African word meaning 'joint-breaker') was identified in a patient in Uganda. In 1960–61 a major epidemic of this *Alphavirus* was identified and the disease spread to Kenya and other countries of East and Central Africa. It was discovered that the virus was transmitted by the *An. gambiae* complex and *An. funestus*. This was the first time an *Anopheles* mosquito was incriminated with the spread of any arbovirus. About 20 other arboviruses infecting humans have since been found to be transmitted by *Anopheles*, some of which are also transmitted by culicine mosquitoes, but a few are known only, or principally, from *Anopheles*.

2.4   CONTROL
The principal methods of mosquito control are described in Chapter 1. Here only those methods specifically directed against *Anopheles*, mainly because of their role as malaria vectors, are considered.

2.4.1   Larval control
Until the availability of DDT and other organochlorines in the mid 1940s, when spraying houses with insecticides became the main strategy in malaria control and malaria eradication programmes, larviciding with petroleum oils and Paris Green was widely carried out to reduce anopheline populations. When properly used, larvicides helped to reduce malaria transmission in localized areas of economic or social importance such as principal towns, coffee and tea estates, rubber plantations and mining camps. However, because of the repetitive nature (every 7–14 days) of this type of control it was logistically impossible to control malaria over large rural areas. This only became possible with the introduction of house-spraying. Nevertheless when larval habitats are readily identified larviciding, such as with temephos, may help reduce anopheline biting. Insect growth regulators such as pyriproxyfen and the microbial insecticide *Bacillus thuringiensis* var. *israelensis* have sometimes been used as part of malaria control campaigns.

Predatory fish, mainly *Gambusia* species, continue to be used to reduce larval populations in some areas, but they have rarely resulted in significantly reduced disease transmission (p. 23).

Drainage of swamps and marshes has in some areas reduced the numbers of *Anopheles*. However, drainage or filling-in of breeding places is impractical in many situations, for example with species that breed in temporary habitats such as scattered pools and puddles, or those breeding in forest pools and swamps.

2.4.2   Adult control

In most malaria control campaigns, control is now focused on the adults.

Residual house-spraying

The most widely practised method is the spraying of water-dispersible (wettable) powders of residual insecticides to the interior surfaces of walls, ceilings and roofs of houses. If vectors have not developed resistance to DDT it remains an effective insecticide, but because DDT accumulates in mammalian tissues and residues have been found in human breast milk many countries have banned it, or are phasing it out. However, presence of residues (which may be minute) does not necessarily mean that human health is affected. A well-recognized disadvantage of DDT is that it may be an irritant to mosquitoes, causing them to leave sprayed houses before they have picked up a lethal dose of insecticide. When used DDT is usually sprayed as a water-dispersible powder at the rate of $2 \, \text{g/m}^2$. Even in highly endemic areas houses need be sprayed only at six-monthly intervals, and where malaria transmission is very seasonal, such as just during the monsoon season, a single spraying before the rains may be sufficient to give good control.

If mosquitoes are resistant to DDT, or there are other reasons why DDT cannot be used, then organophosphates such as malathion, fenthion or fenitrothion or the carbamate propoxur can be used, usually at rates of $1.5–2 \, \text{g/m}^2$. Bendiocarb, another carbamate, has a lower dosage rate of just $0.4 \, \text{g/m}^2$. Alternatively, lambda-cyhalothrin at $0.02–0.03 \, \text{g/m}^2$ or permethrin at $0.5 \, \text{g/m}^2$ can be used. However, all these alternative insecticides are less persistent and spraying may have to be repeated every 3–4 months. They are also more expensive than DDT, but sometimes the lower dosage rates compensate for this. Most are also more toxic to people than DDT: consequently stricter safety measures must be introduced during control programmes.

House-spraying often becomes popular because it also kills bedbugs, house-flies and cockroaches. However, the effectiveness of spraying houses in malaria control programmes depends on the mosquitoes resting indoors, and many important vectors, especially those in South-east Asia and the tropical Americas, feed and/or rest out of doors, and are thus not killed by residual house-spraying. There is also evidence that house-spraying may promote the selection of exophilic populations of a species, that is populations that before spraying rested in houses now rest out of doors. Similarly, spraying may reduce populations of one species of a complex that is primarily endophilic, but by so doing allow an increase in numbers of another species in the complex that is exophilic. Such population changes have been recorded in the *An. gambiae* complex in Zimbabwe, *An. nunez-tovari* in Venezuela, *An. minimus* in Thailand and in a few other species.

Insecticide-impregnated bed-nets

Many of the bed-nets bought by poor communities are cheap, badly made and soon tear, and torn nets are useless in protecting people against mosquito bites. Also, mosquitoes will readily bite through a net if any part of the body is pressed up against it. However, nets that are impregnated with insecticides such as permethrin (200–500 mg/m^2), deltamethrin (15–25 mg/m^2) or lambda-cyhalothrin (10–20 mg/m^2), all of which repel and kill mosquitoes, will still protect people against bites even if they are torn or people sleep up against them. Nets can be impregnated by simply dipping them in insecticide contained in a plastic dustbin. Such impregnated nets should remain effective for 4–6 months before they need re-impregnating. If, however, lambda-cyhalothrin is used for impregnation nets can remain effective for a year or more. Washing nets usually makes them less effective. The latest development involves impregnating the fibres used to make nets with permethrin during their manufacture. Nets made with insecticide-impregnated fibres apparently remain effective for 4 years.

In Vietnam and in China, where there are more than 3.85 million treated nets, there have been significant reductions in malaria transmission and morbidity. In West Africa and Kenya 'all-cause' child mortality in large-scale trials has been reduced by 17–36%. However, reductions in morbidity and mortality have generally been variable in Africa. It appears that if used on a large scale impregnated nets can have a mass-killing effect on vector populations which can benefit the whole community, even those without nets. Many regard such nets as the main hope for malaria control; however, the incidence of resistance to pyrethroids is slowly increasing.

Nets will also give protection against filariasis vectors, but will not be effective against any vector that bites out of doors or in the early evening before people have gone to bed.

2.4.3   Malaria control and malaria eradication

In 1955 the eighth World Health Assembly stated that worldwide malaria eradication, except in Africa south of the Sahara, was technically feasible. However, a sense of urgency in achieving this aim was recognized because insecticide resistance in *Anopheles* had been reported in 1950. In

1968 the 22nd World Health Assembly realized it had been overoptimistic and declared that global malaria eradication was not at present possible, although it remained the ultimate goal, and that for the time being malaria control should be the aim. This basically remains the situation today. The latest initiative, announced in 1998, is a joint WHO, UNDP, UNICEF and World Bank venture called the global Roll Back Malaria (RBM) programme, which will initially concentrate on Africa. It advocates an integrated approach to malaria control, involving impregnated mosquito nets, community level drug treatment, and vaccines. The goal is a 50% reduction in malaria mortality by the year 2010.

The difference between malaria eradication and malaria control is that malaria eradication means the total cessation of transmission and elimination of the reservoir of infection in people so that at the end of the anti-malaria campaign there is no resumption of transmission. Malaria control means reducing malaria transmission to an acceptable rate, that is to a level that no longer constitutes a major public health problem. This has the implication that control measures have to be maintained indefinitely; if they are relaxed malaria prevalence will rise. The feasibility of control will depend not only on scientific considerations but also on the financial and public health resources of the community, or country.

FURTHER READING

Beier, J. C. (1998) Malaria parasite development in mosquitoes. *Annual Review of Entomology*, **43**, 519–43.

Chavasse, D., Reed, C. and Attawell, K. (eds.) (1999) *Insecticide Treated Net Projects: a Handbook for Managers*. London and Liverpool: Malaria Consortium.

Collins, F. H. and Paskewitz, S. M. (1995) Malaria: current and future prospects for control. *Annual Review of Entomology*, **40**, 195–219.

Coluzzi, M. and Bradley, D. (1999) The malaria challenge after one hundred years of malariology. *Parassitologia*, **41** (1–3), 1–528.

Curtis, C. F. and Townson, H. (1998) Malaria: existing methods of vector control and molecular entomology. *British Medical Bulletin*, **54**, 311–25.

Dean, M. (2001) *Lymphatic Filariasis: the Quest to Eliminate a 4000-Year-Old Disease*. Hollis, NH: Hollis Publishing Company.

Dobson, M., Malowany, M. and Stapleton, D. (eds.) (2000) Dealing with malaria in the last sixty years. Proceedings of a Rockefeller Foundation Conference held in New York, May 11–14, 1998. *Parassitologia*, **42** (1–2), 1–182.

Gilles, H. M. and Warrell, D. A. (eds.) (2002) *Essential Malariology*. 4th edn. London: Edward Arnold.

Litsios, S. (1996) *The Tomorrow of Malaria*. Wellington, New Zealand: Pacific Press.

Molyneux, L. and Gramiccia, G. (1980) *The Garki Project: Research on the Epidemiology and Control of Malaria in the Sudan Savanna of West Africa*. Geneva: World Health Organization.

Nájera, J. A. (2001) Malaria control: achievements, problems and strategies. *Parassitologia*, **43** (1–2), 1–98.

Nájera, J. A., Kouznetsov, R. L. and Delacollette, C. (1998) *Malaria Epidemics: Detection and Control, Forecasting and Prevention*. WHO/MAL/98.1084. Geneva: World Health Organization.

Rozendaal, J. A. (1989) Impregnated mosquito nets and curtains for self-protection and vector control. *Tropical Disease Bulletin*, **86**, R1–R41.

Sasa, M. (1976) *Human Filariasis: a Global Survey of Epidemiology and Control*. Tokyo: University of Tokyo Press.

Wernsdorfer, W. H. and McGregor, I. (eds.) (1988) *Malaria: Principles and Practice of Malariology*. Volumes 1 and 2. Edinburgh: Churchill Livingstone.

World Health Organization (1989) *Geographical Distribution of Arthropod-Borne Diseases and their Principal Vectors*. WHO/VBC/89.967. Geneva: World Health Organization Vector Biology and Control Division.

(1992) Lymphatic filariasis: the disease and its control. *World Health Organization Technical Report Series*, **821**, 1–71.

(1992) *Entomological Field Techniques for Malaria Control. Part 1: Learner's Guide*. Geneva: World Health Organization.

(1992) *Entomological Field Techniques for Malaria Control. Part II: Tutor's Guide*. Geneva: World Health Organization.

(1993) *A Global Strategy for Malaria Control*. Geneva: World Health Organization.

(1993) Implementation of the global malaria control strategy. *World Health Organization Technical Report Series*, **839**, 1–57.

(1994) *Lymphatic Filariasis Infection and Disease: Control Strategies*. Report of a consultative meeting at the Universitii Sains Malaysia Penang, Malaysia (August 1994). TDR/CTD/FIL/PENANG/94.1, 1–30.

(1995) Vector control for malaria and other mosquito-borne diseases. *World Health Organization Technical Report Series*, **857**, 1–91.

Zahar, A. R. (1984–96) A series of ten World Health Organization mimeographed documents entitled 'Vector bionomics in the epidemiology and control of malaria', covering Europe, Africa, the southern and eastern Mediterranean regions, south-western Arabia, Asia west of India, South-east Asia and the western Pacific region.

See also references at the ends of Chapters 1 and 3.

3

Culicine mosquitoes (Culicinae)

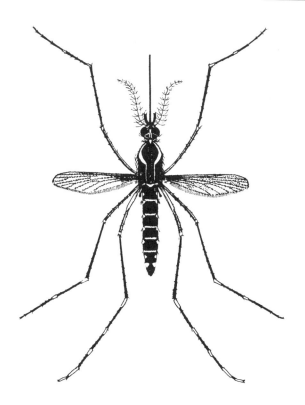

The subfamily Culicinae contains 37 genera of mosquitoes, of which the medically most important ones are *Culex, Aedes, Ochlerotatus, Psorophora, Mansonia, Haemagogus* and *Sabethes*. The genera *Culex, Aedes, Ochlerotatus* and *Mansonia* are found in both temperate and tropical regions, whereas *Psorophora* species are found only in North, Central and South America. *Haemagogus* and *Sabethes* mosquitoes are restricted to Central and South America.

Certain *Aedes* mosquitoes are vectors of yellow fever in Africa, and *Aedes, Haemagogus* and *Sabethes* are yellow fever vectors in Central and South America. *Aedes* species are also vectors of the classical and haemorrhagic forms of dengue. All seven genera of culicine mosquitoes mentioned here, as well as some others, can transmit a variety of other arboviruses. Some *Culex, Aedes, Ochlerotatus* and *Mansonia* species are important vectors of filariasis (*Wuchereria bancrofti* or *Brugia malayi*). *Psorophora* species are mainly pest mosquitoes but a few transmit arboviruses.

Characters separating the subfamily Culicinae from the Anophelinae have been given in Chapter 1 and are summarized in Table 1.1.

It is not easy to give a reliable and non-technical guide to the identification of the most important culicine genera. Nevertheless, characters that will usually separate these genera are given below, together with notes on their biology.

3.1   *CULEX* MOSQUITOES

3.1.1   Distribution
Culex mosquitoes are found more or less worldwide, but they are absent from the extreme northern parts of the temperate zones.

3.1.2   Eggs
Eggs are usually brown, long and cylindrical, laid upright on the water surface and placed together to form an *egg raft* which can comprise up to about 300 eggs (Fig. 1.15). No glue or cement-like substance binds the eggs to each other; adhesion is due to surface forces holding the eggs together. Eggs of a few other mosquitoes, including those of the genus *Coquillettidia*, also deposit their eggs in rafts.

3.1.3   Larvae
The larval siphon is often long and narrow (Fig. 3.1), but it may be short and fat. There is always *more than one pair* of *subventral tufts* of hairs on the siphon, none of which is near its base. These hair tufts may consist of very few short and simple hairs which may be missed unless larvae are carefully examined under a microscope.

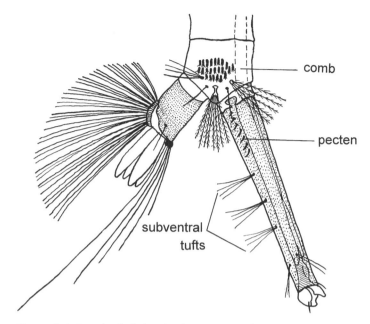

Figure 3.1 Terminal abdominal segments of a *Culex* larva showing the long siphon with more that one (three shown here) subventral tufts of hair.

3.1.4 Adults

Frequently, but not always, the thorax, legs and wing veins of the adult are covered with ***dull-coloured***, often brown, scales. The abdomen is often covered with brown or blackish scales but some whitish scales may occur on most segments. Adults are recognized more by their lack of ornamentation than by any striking diagnostic characters. The tip of the female abdomen is blunt. Claws on all tarsi are simple and those on the hind tarsi are very small. Examination under a microscope shows that all tarsi have a pair of small fleshy ***pulvilli*** (Fig. 1.2).

3.1.5 Biology

Eggs are laid in a great variety of aquatic habitats. Most *Culex* species breed in ***ground collections*** of water such as pools, puddles, ditches, borrow pits and ricefields. Some lay eggs in man-made container-habitats such as tin cans, water receptacles, bottles and storage tanks. Only a few species breed in tree-holes and even fewer in leaf axils. The medically most important species, *Culex quinquefasciatus*, which is a vector of bancroftian filariasis, breeds in waters polluted with organic debris such as rotting vegetation, household refuse and excreta. Larvae of this species are commonly found in partially blocked drains and ditches, soak-away pits and septic tanks, and in village pots, especially the abandoned ones in which water is polluted and unfit for drinking. It is a mosquito that is associated with urbanization, and towns with poor and inadequate drainage and sanitation. Under these conditions its population increases rapidly. Adults mainly bite at night.

Culex tritaeniorhynchus is an important vector of Japanese encephalitis and breeds prolifically in ricefields and also in grassy pools. In southern Asia larvae are not uncommon in fish ponds which have had manure added to them.

Culex quinquefasciatus, and many other *Culex* species, bite humans and other hosts ***at night***. Some species, such as *C. quinquefasciatus*, commonly rest indoors both before and after feeding, but they also shelter in outdoor resting places.

3.2   *AEDES* AND *OCHLEROTATUS* MOSQUITOES

In 2000 the subgenus *Ochlerotatus* of the genus *Aedes* was raised to generic status. This means that several medically important mosquitoes formerly in the genus *Aedes* are now in the genus *Ochlerotatus*. This will inevitably cause confusion to non-mosquito experts! However, the very minor morphological differences separating *Aedes* and *Ochlerotatus* need not trouble the reader, and moreover the biology of the species in the two genera is virtually the same, so these two genera are treated together in this account. In practice all the change means is that some of the mosquitoes formerly called *Aedes* are know called *Ochlerotatus*: for example *Aedes togoi* has become *Ochlerotatus togoi*.

3.2.1   Distribution

Worldwide, the range of *Aedes* and *Ochlerotatus* mosquitoes extends well into northern and Arctic areas, where they can be vicious biters and serious pests to people and livestock.

3.2.2   Eggs

Eggs are usually black, more or less ovoid in shape, and are always laid *singly* (Fig. 1.15). Careful examination shows that the eggshell has a distinctive mosaic pattern. Eggs are laid on damp substrates just beyond the water line, such as on damp mud and leaf litter of pools, on the damp walls of clay pots, rock-pools and tree-holes.

Aedes and *Ochlerotatus* eggs can withstand ***desiccation***, the intensity and duration of which varies, but in many species they can remain dry but viable for many months. When flooded, some eggs may hatch within a few minutes, while others of the same batch may require prolonged immersion in water; thus hatching may be spread over several days or weeks. Even when eggs are soaked for long periods some may fail to hatch because they require several soakings followed by short periods of desiccation before hatching can be induced. Even if environmental conditions are favourable, eggs may be in a state of diapause and will not hatch until this resting

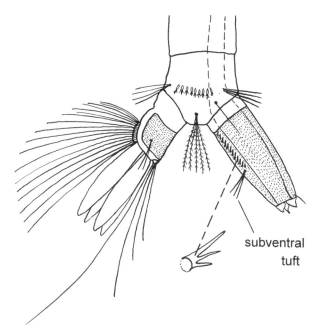

Figure 3.2 Terminal abdominal segments of an *Aedes* or *Ochlerotatus* larva showing the short siphon with a single subventral hair tuft.

period is terminated. Various stimuli, including reduction in the oxygen content of water, changes in daylength, and temperature, may be required to break diapause in *Aedes* and *Ochlerotatus* eggs.

Many *Aedes* species, but fewer *Ochlerotatus* species, breed in small *container-habitats* (tree-holes, plant axils, etc.) which are susceptible to drying out; thus the ability of eggs to withstand *desiccation* is clearly advantageous. Desiccation and the ability of *Aedes* and *Ochlerotatus* eggs to hatch in instalments can create problems with controlling the immature stages (p. 74).

3.2.3 Larvae

Aedes and *Ochlerotatus* species usually have a short barrel-shaped siphon, and there is only *one pair* of *subventral tufts* (Fig. 3.2) which never arises from less than one-quarter of the distance from the base of the siphon. Additional characters are at least three pairs of setae in the ventral brush, antennae that are not greatly flattened and no enormous setae on the thorax. These characters should separate *Aedes* and *Ochlerotatus* larvae from most of the culicine genera, but not unfortunately from larvae of South American *Haemagogus*. In Central and South America *Aedes* and *Ochlerotatus* larvae can usually be distinguished from those of *Haemagogus* by possessing either

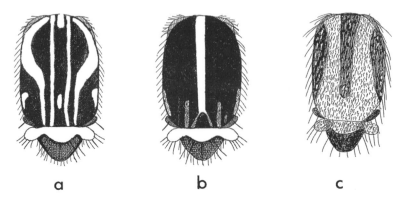

a   b   c

Figure 3.3 Dorsal surfaces of the thoraces of adult *Aedes* and *Ochlerotatus* mosquitoes showing examples of thoracic patterns of dark and pale scales: (a) *Aedes aegypti* with black scales and typical lyre-shaped silvery markings; (b) *Aedes albopictus* with prominent silvery central line; (c) *Ochlerotatus trivittatus* with dark brown and whitish, not silvery, scales.

larger or more strongly spiculate antennae; also the comb is not on a sclerotized plate as in some *Haemagogus*.

3.2.4   Adults

Many, but not all, *Aedes* adults have conspicuous **patterns** on the thorax formed by black, white or silvery scales (Fig. 3.3a,b); in some species yellow and/or brownish scales are present. Many *Ochlerotatus* species also have conspicuous thoracic patterns (Fig. 3.3c), but the pale scales are usually white, not silvery, and dark scales are often more brown than black. The legs of both *Aedes* and *Ochlerotatus* species often have dark (often black) and white rings (Fig. 3.4b). *Aedes aegypti*, often called the yellow fever mosquito, is readily recognized by the lyre-shaped silver markings on the lateral edges of the scutum (Fig. 3.3a). Scales on the wing veins of *Aedes* and *Ochlerotatus* mosquitoes are narrow and usually mainly black. In *Aedes* and *Ochlerotatus* the abdomen is often covered with black and white scales forming distinctive patterns, and in the female it is pointed at its tip (Fig. 3.4a).

3.2.5   Biology

Although some *Aedes* and *Ochlerotatus* species breed in marshes and ground pools, including snow-melt pools in arctic and subarctic areas, many, especially tropical species, are found in natural **container-habitats** such as tree-holes, bamboo stumps, leaf axils and rock-pools, or in man-made ones such as village pots, tin cans and tyres. For example, *Ae. aegypti* breeds in village pots and water storage jars placed either inside or outside

Figure 3.4 (a) Abdomen and (b) leg of an *Aedes* or *Ochlerotatus* adult showing typical arrangement of black and white scales.

houses. Larvae occur mainly in those with clean water intended for drinking. In some areas *Ae. aegypti* also breeds in rock-pools and tree-holes. *Aedes africanus*, an African species involved in the sylvatic transmission of yellow fever, breeds mainly in tree-holes and bamboo stumps, whereas *Ae. bromeliae*, another African yellow fever vector, breeds almost exclusively in leaf axils, especially those of banana plants, pineapples and coco-yams (*Colocasia*).

Aedes albopictus, a vector of dengue in South-east Asia, breeds in both natural and man-made container-habitats such as tree-holes, water pots and vehicle tyres. This species was introduced into the USA in 1985 as dry but viable eggs which had been oviposited in tyres in Asia and then exported. By 2003 it had spread to 28 states in the USA, including Hawaii, and had also been introduced by export of tyres to Argentina, Brazil, possibly Bolivia, the Cayman Islands, Colombia, Cuba, the Dominican Republic, El Salvador, Guatemala, Mexico, Panama, Fiji, Papua New Guinea, Cameroon, Nigeria, Bioko (Equatorial Guinea), Albania, France, Italy and Israel.

Larvae of *Ae. polynesiensis* occur in man-made and natural containers, especially split coconut shells, whereas larvae of *Ae. pseudoscutellaris* are found in tree-holes and bamboo stumps. Both species are important vectors of diurnal subperiodic bancroftian filariasis. *Ochlerotatus togoi*, a minor vector of nocturnal periodic bancroftian and brugian filariasis, breeds principally in rock-pools containing fresh or brackish water.

The life cycle of *Aedes* and *Ochlerotatus* mosquitoes from eggs to adults can be rapid, taking as little as about 7 days, but it more usually takes 10–12 days; in temperate species the life cycle may last several weeks to many months, and some species overwinter as eggs or larvae.

Adults of most *Aedes* and *Ochlerotatus* species bite mainly during the *day* or early evening. Most biting occurs out of doors and adults usually rest out of doors before and after feeding.

3.3   *HAEMAGOGUS* MOSQUITOES

3.3.1   Distribution
Haemagogus mosquitoes are found only in Central and South America.

3.3.2   Eggs
The eggs are usually black and ovoid and laid singly in tree-holes and other natural container-habitats, occasionally in man-made ones. There is no simple method of distinguishing eggs of *Haemagogus* from those of *Aedes*, *Ochlerotatus* or *Psorophora* mosquitoes.

3.3.3   Larvae
Larvae have only one pair of subventral tufts arising, as in *Aedes* and *Ochlerotatus*, not less than a quarter of the distance from the base of the siphon. They resemble *Aedes* and *Ochlerotatus* larvae but can usually be distinguished by the antennae being short and either without spicules or with just a very few, and by a ventral brush arising from a sclerotized *boss* (Fig. 3.5). In some species the comb teeth are arranged at the edge of a sclerotized *plate*; in *Aedes* and *Ochlerotatus* this plate is absent.

3.3.4   Adults
Adults are very *colourful* and can easily be recognized by the presence of broad, flat and bright metallic blue, red, green or golden scales, covering the dorsal part of the thorax. Like *Sabethes* mosquitoes (Fig. 3.7a) they have exceptionally large *antepronotal thoracic lobes* behind the head. *Haemagogus* adults are rather similar to *Sabethes* in other respects, and it may be difficult for the novice to separate these two genera. However, no *Haemagogus* mosquito has paddles on the legs, which is a conspicuous feature of many, but not all, species of *Sabethes* (Fig. 3.7c).

3.3.5   Biology
Eggs can withstand *desiccation*. Larvae occur mostly in tree-holes and bamboo stumps, but also in rock-pools, split coconut shells and some-times in assorted domestic containers. They are basically forest mosquitoes. Adults bite during the day but mostly in the tree-tops where they feed on monkeys. Under certain environmental conditions, however, such as

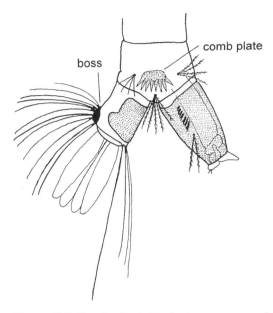

Figure 3.5 Terminal abdominal segments of a *Haemagogus* larva showing a ventral brush arising from a dark sclerotized boss, and comb scales arranged on a small 'plate'.

experienced at edges of forests during tree-felling operations or during the dry season, they may descend to the forest floor to bite humans and other hosts. Species such as *Haemagogus spegazzinii*, *Hg. leucocelaenus* and *Hg. janthinomys* are all involved in yellow fever transmission in forested areas.

3.4   *SABETHES* MOSQUITOES

3.4.1   Distribution
Sabethes mosquitoes are found only in Central and South America.

3.4.2   Eggs
Little is known about the eggs of *Sabethes* species, but it appears that they are laid singly, have no prominent surface features such as bosses or sculpturing and are incapable of withstanding desiccation. The eggs of *Sabethes chloropterus*, a species sometimes involved in the sylvatic cycle of yellow fever, are rhomboid in shape and can thus be readily identified from most other culicine eggs (Fig. 3.7b).

3.4.3   Larvae
The larval siphon is relatively slender and moderately long (Fig. 3.6) and has many *single* hairs placed ventrally, laterally or dorsally. *Sabethes* larvae

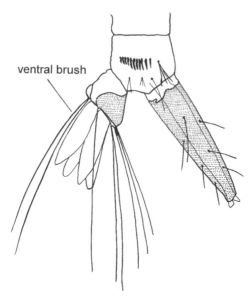

Figure 3.6 Terminal abdominal segments of a *Sabethes* larva showing a single pair of hairs in the ventral brush, numerous single hairs on the siphon, and absence of a pecten on the siphon.

can usually be distinguished from other mosquito larvae by having only *one pair* of setae in the ventral brush, the comb teeth arranged in a single row, or at most with three or four detached teeth, and by the absence of a pecten.

3.4.4 Adults

The dorsal surface of the thorax of the adult is covered with appressed *iridescent* blue, green and red scales. The *antepronotal lobes*, like those in *Haemagogus*, are very large (Fig. 3.7a). Adults of many species have one or more pairs of tarsi with conspicuous *paddles* composed of narrow scales (Fig. 3.7c). Their presence immediately distinguishes *Sabethes* from all other mosquitoes. Species which lack these paddles resemble those of *Haemagogus* and a specialist is required to distinguish them.

3.4.5 Biology

Larvae occur in tree-holes and bamboo stumps; a few species are found in leaf axils of bromeliads and other plants. They are forest mosquitoes. Adults bite during the day and mainly in the tree canopy, but like *Haemagogus* adults may descend to ground level at certain times to bite humans and other hosts. *Sabethes chloropterus* has been incriminated as a sylvan vector of yellow fever.

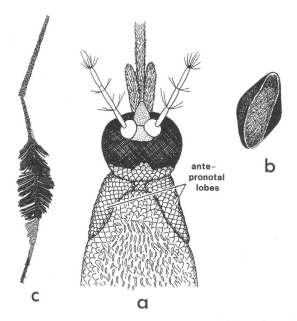

ante-
pronotal
lobes

b

c

a

Figure 3.7 *Sabethes* mosquitoes: (a) head and thorax showing antepronotal lobes forming a 'collar' behind the head; (b) an egg; (c) hind leg showing long narrow scales forming a 'paddle'.

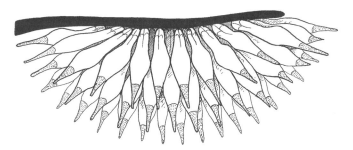

Figure 3.8 *Mansonia* eggs glued to the underside of floating aquatic vegetation.

3.5   *MANSONIA* MOSQUITOES

3.5.1   Distribution

Mansonia is principally a genus of wet tropical areas, but a few species occur in temperate regions.

3.5.2   Eggs

Eggs are dark brown or blackish and cylindrical, but have a tube-like extension apically which is usually darker than the rest of the egg. Eggs are laid in sticky compact *masses*, often arranged as a rosette, which are glued to the undersurface of floating vegetation (Figs. 1.15, 3.8).

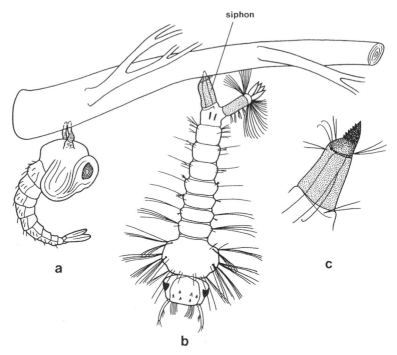

Figure 3.9 Immature stages of *Mansonia* mosquitoes: (a) pupa with respiratory trumpets inserted into an aquatic plant; (b) larva with siphon inserted into an aquatic plant for respiration; (c) larval siphon showing serrated structures used to pierce aquatic plants.

3.5.3   Larvae

Mansonia larvae are easily recognized because they have ***specialized siphons*** adapted for piercing aquatic plants (Fig. 3.9b,c) to obtain air. The siphon tends to be conical with the apical part darker and heavily sclerotized, and it has teeth and curved hairs which enable the larva to attach to plants and insert its siphon. The pupal respiratory trumpets are also inserted into plants for respiration (Fig. 3.9a).

3.5.4   Adults

Typically adults have the legs (Fig. 3.10c), palps, wings and body covered with a ***mixture*** of dark (usually brown) and pale (usually white or creamy) scales, giving the mosquito a rather dusty appearance. The scattering of dark and pale scales on the wing veins gives the wings the appearance of having been sprinkled with salt and pepper (Fig. 3.10a), and provides a useful character for identification. Closer examination shows that the scales on the wings are very broad and often asymmetrical and almost ***heart-shaped*** (Fig. 3.10b). In other mosquitoes these scales are longer and narrower.

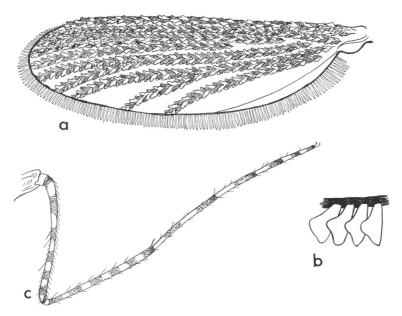

Figure 3.10 *Mansonia* mosquitoes: (a) wing showing speckled distribution of dark and pale scales on the wings; (b) a few scales on a
wing vein showing their broad, almost heart shape; (c) leg showing
distribution of dark and pale scales giving a banded pattern.

3.5.5   Biology

Eggs are glued to the undersurface of plants and hatch within a few days;
they are unable to withstand desiccation. All larval habitats have aquatic
vegetation, either rooted (such as grasses, rushes and reeds) or floating
(such as *Pistia stratiotes*, *Salvinia* or *Eichhornia* species). Larvae consequently
occur in permanent collections of waters, such as swamps, marshes, ponds,
borrow pits, grassy ditches, irrigation canals and even in the middle of
rivers if they have floating plants.

Larvae and pupae only detach themselves from plants and rise to the
surface of the water if they are disturbed. Because they are more or less
permanently attached to plants the immature stages are frequently missed
in larval surveys. It is therefore not easy to identify breeding places unless
special collecting procedures are undertaken, such as the removal of plants
and their examination for attached *Mansonia* larvae and pupae. Because of
problems of getting insecticides to the larvae, which may be some distance
below the water surface, it is often difficult to control breeding of *Mansonia*
species with conventional insecticidal applications, although herbicides
can be used to kill the plants (see p. 76).

Adults usually bite at night, but a few species are day-biters. After feeding, most *Mansonia* rest out of doors, but a few species rest indoors. The

main medical importance of *Mansonia* mosquitoes is as vectors of filariasis, such as the nocturnal periodic and nocturnal subperiodic forms of *Brugia malayi* in Asia. In Africa filariasis (*W. bancrofti*) is not transmitted by *Mansonia* species, although they can be vectors of a few, but not very important, arboviruses.

3.6   *COQUILLETTIDIA* MOSQUITOES

The genus *Coquillettidia* is of lesser medical importance and therefore described only briefly. It is mainly tropical but a few species occur in temperate regions. *Coquillettidia* is related to *Mansonia* and has sometimes been treated as a subgenus of *Mansonia*, but species of *Coquillettidia* differ in several respects. For example, as in *Culex* species, eggs are formed into *egg rafts* that float on the water, but they are narrower and longer than *Culex* rafts. Although larvae have rather conical siphons, which like those of *Mansonia* are inserted into plants for respiratory purposes, the larval antennae are much longer than those in *Mansonia*. Adults have narrow scales on the wings, not broad or heart-shaped ones as found in *Mansonia*, and several *Coquillettidia* species are a bright *yellow*. *Coquillettidia crassipes* can be a vector of nocturnal subperiodic *B. malayi* in South-east Asia.

3.7   *PSOROPHORA* MOSQUITOES

Psorophora mosquitoes are of lesser medical importance and so are described briefly. They are found from Canada to South America. They are similar in many respects to *Aedes* and *Ochlerotatus* species: for example their eggs look like eggs of these two genera and like them they can withstand *desiccation*. Adults of some pest species are large mosquitoes. A specialist is required to distinguish the larvae and adults of the three genera. Breeding places are mainly flooded pastures and sometimes ricefields; larvae of several species are predators. Although they can be vectors of a few arboviruses, such as Venezuelan equine encephalitis, their main importance is as vicious biters.

3.8   MEDICAL IMPORTANCE

3.8.1   Biting nuisance

In several areas of the world a lot of money is spent on mosquito control, not so much because mosquitoes are vectors of disease but because they are troublesome biters. For example, some of the best-organized mosquito control operations are in North America, where more money is spent on killing mosquitoes than is expended in most tropical countries where they are important vectors of disease. In northern temperate and subarctic parts of America, Europe and Asia much greater numbers of *Aedes* and *Ochlerotatus* mosquitoes can be encountered biting people than in tropical countries. Although they may not be transmitting diseases to humans in these areas they can, nevertheless, make life outdoors almost intolerable.

3.8.2   Arboviruses

Numerous arboviruses are transmitted by culicine mosquitoes, including important ones such as those causing yellow fever and dengue fever.

The intrinsic incubation period of a virus in humans is usually just a few days, often 3–4 days, after which virus appears in the peripheral blood and the host is viraemic. *Viraemia* lasts only a few days, typically 3 days, after which the virus disappears from the peripheral blood. A vector must bite a viraemic host if it is to become infected and transmit an infection.

A relatively high titre of arbovirus, however, is usually required before virus can pass across the gut cells of the mosquito into the haemolymph. From the haemolymph the virus invades many tissues and organs, including the *salivary glands*, where virus multiplication occurs. This is the *extrinsic cycle* of development and can take 5–30 days, depending on temperature, the type of virus and the mosquito species, but in most mosquito-borne viral infections the extrinsic cycle is typically 10–15 days.

3.8.3   Yellow fever (*Flavivirus*)

The arbovirus causing yellow fever occurs in Africa and tropical areas of the Americas. It does not occur in Asia or elsewhere, although mosquitoes capable of transmitting the disease occur in many countries. Yellow fever is a *zoonosis*, being essentially an infection of forest monkeys which under certain conditions can be transmitted to humans.

Africa

In Africa the yellow fever virus occurs in certain cercopithecid monkeys inhabiting the forests and is transmitted amongst them by tree-hole-breeding mosquitoes, mainly *Aedes africanus*. This forest-dwelling mosquito bites mainly in the forest canopy soon after sunset – just in the right place at the right time to bite monkeys going to sleep in the tree-tops. This sylvatic, forest or monkey cycle, as it is sometimes called, maintains a virus reservoir in the monkey population (Fig. 3.11). In Africa, monkeys are little affected by yellow fever, dying only occasionally. Some of the monkeys involved in the forest cycle descend from the trees to steal bananas from farms at the edge of the forest. In this habitat the monkeys get bitten by different mosquitoes. For example, by *Aedes bromeliae* (formerly called *Ae. simpsoni*), a species that breeds in leaf axils of bananas, plantains, coco-yams (*Colocasia* species) and pineapples, and in West Africa also by other species such as tree-hole-breeding *Aedes furcifer*, *Ae. taylori* and *Ae. luteocephalus*. These species bite during the day at the edges of forests and if the monkeys have viraemia, that is yellow fever virus circulating in their peripheral blood, the mosquitoes become infected. If these mosquitoes live sufficiently long they can transmit yellow fever to other monkeys, or more importantly to people. This transmission cycle, occurring in clearings at the edge of the forest involving monkeys and humans, is sometimes referred to as the rural cycle (Fig. 3.11). When people return to their villages, or travel

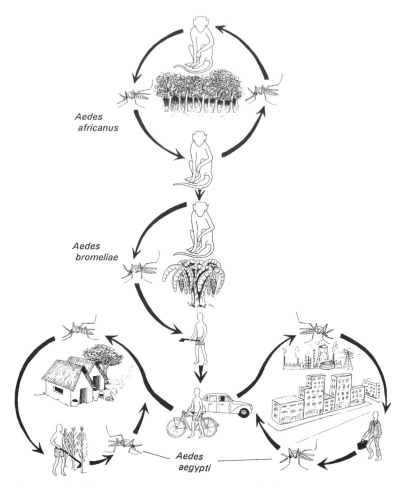

Figure 3.11 Diagrammatic representation of the sylvatic, rural and urban cycles of yellow fever transmission in Africa (only major vectors included).

to towns, they get bitten by different mosquitoes, including *Ae. aegypti*, a domestic species breeding mainly in man-made containers such as water storage pots, abandoned tin cans and vehicle tyres. If people have viraemia then *Ae. aegypti* becomes infected and yellow fever is transmitted among the human population by this species. This is the urban cycle of yellow fever transmission (Fig. 3.11).

The epidemiology of yellow fever is complicated and variable. In some areas, for example, yellow fever may be circulating among the monkey population yet rarely gets transmitted to humans because local vector mosquitoes are predominantly zoophagic. Other primates in Africa such as bush-babies (*Galago* species) may also be reservoirs of yellow fever.

Figure 3.12 Diagrammatic representation of the jungle, rural and urban cycles of yellow fever transmission in Central and South America.

Yellow fever virus may be ***transovarially*** transmitted in *Aedes* species. This means the eggs and subsequent generation of adults are born already infected with virus. (Transovarial transmission is more usually associated with tick-borne diseases than with mosquitoes.) The virus can also be passed from congenitally infected males to females during mating, which in effect is venereal transmission. How important these routes of infection are remain unclear, but such mechanisms, especially transovarial transmission, would provide a means of virus survival during long dry seasons.

Americas

In Central and South America the yellow fever cycle, although similar to that in Africa, differs in certain aspects (Fig. 3.12). As in Africa it is an infection of forest monkeys, mainly cebid ones (e.g. howler and spider monkeys), and is transmitted among them by forest-dwelling mosquitoes. The jungle vectors are various *Haemagogus* species including *Hg. spegazzinii*, *Hg. janthinomys* and *Hg. leucocelaenus*. *Sabethes chloropterus* is also a vector although it seems to be a relatively inefficient one, and sometimes *Aedes* species are also involved in transmission. These are all arboreal mosquitoes which bite in the forest canopy and breed in tree-holes. New World monkeys are more susceptible to yellow fever than African monkeys and they frequently become sick and die. When people enter the jungle to cut down trees for timber, mosquitoes, which normally bite monkeys at canopy heights, may descend and bite them; if these mosquitoes are infected people will develop yellow fever. The disease is then spread from person to person in villages and towns, as in Africa, by *Ae. aegypti* (Fig. 3.12). It is possible that in South America marsupials and even sloths and porcupines may be involved in yellow fever cycles.

In Panama **transovarial** transmission has been reported in *Haemagogus* species.

3.8.4 Dengue (*Flavivirus*)

A number of different viruses (dengue types 1, 2, 3 and 4) are responsible for dengue. Dengue is widely distributed in the tropics, occurring throughout most of South-east Asia, the Pacific, the Indian subcontinent, Africa, the USA down to northern parts of South America, and in the Caribbean. A more severe form, dengue haemorrhagic fever, causes infant mortality and has appeared in many parts of South-east Asia and also India. In 1981 haemorrhagic dengue and dengue type 4 were first noticed in Cuba. Dengue type 4 and haemorrhagic dengue have since been reported from at least 32 countries in the Americas, including many of the Caribbean islands. Dengue is now the most important viral disease transmitted by mosquitoes, having been recorded from more than 100 countries, and the numbers of cases worldwide are increasing.

Both dengue and haemorrhagic dengue are transmitted by *Ae. aegypti* and in South-east Asia to a lesser extent also by *Ae. albopictus*; both species breed in natural and man-made container-habitats such as water pots and tyres. Mosquitoes of the *Ae. scutellaris* group, which also breed in natural and man-made containers, may also be vectors in the Pacific islands. Although transmission of dengue virus amongst monkeys in forests of Malaysia, Vietnam and West Africa has been reported there is no evidence that enzootic strains are involved in epidemics.

3.8.5 Other arboviruses

There are many other mosquito-borne arboviruses infecting humans in various parts of the world, for example Ross River virus (*Alphavirus*: Australasia), Chikungunya (*Alphavirus*: Africa, Asia), Sindbis (*Alphavirus*:

Africa, Asia, Australia, Europe), Murray Valley encephalitis (*Alphavirus*: Australia, Papua New Guinea) and Rift Valley fever (*Bunyavirus*: Africa, Saudi Arabia, Yemen). These and the more important arboviral diseases very briefly discussed below are *zoonoses*, as of course is yellow fever.

West Nile virus (WN) (*Flavivirus*). In Africa, Europe, Middle East, and in 1999 first recorded from the Americas, in New York, and later in Canada. How the virus entered North America is unresolved, but by 2003 it had been recorded in 44 US states, five Canadian provinces and in Mexico. In 2002 antibodies apparently to WN virus were found in birds in the UK. Migratory birds are suspected of spreading the virus and are the main reservoir hosts; principal vectors are *Culex* species such as *Cx. pipiens* and also ticks (e.g. *Argas* and *Hyalomma* species). Horses and camels also become infected. Human infections are also acquired from *Culex* mosquitoes such as *Cx. pipiens, Cx. modestus* and *Culex univittatus*, species which feed also on birds.

Japanese encephalitis (JE) (*Flavivirus*). In Japan, China, Malaysia, Korea and other areas of South-east Asia, India and a very few cases in Australia. The basic transmission cycle involves water birds, mainly herons, egrets and ibises. These birds and also pigs are important reservoir hosts (*maintenance hosts*) and both, but especially pigs, develop high viraemias and are called *amplifying hosts*. Transmission to birds, humans and pigs is mainly by *Cx. tritaeniorhynchus*, which is a common ricefield breeding mosquito. *Culex gelidus*, breeding in streams and ricefields, is another vector and probably maintains the virus in pig-to-pig transmissions in South-east Asia, while the *Cx. vishnui* group are vectors in India.

St Louis encephalitis (SLE) (*Flavivirus*). In southern Canada and widely distributed in the USA, extending into Central and South America. In eastern USA it is mainly an urban disease spread by *Cx. quinquefasciatus* or *Cx. pipiens*. *Amplifying hosts* are chickens and peridomestic wild birds. In western USA SLE occurs mainly in rural areas with the ricefield-breeding *Cx. tarsalis* as a vector.

Eastern equine encephalitis (EEE) (*Alphavirus*). In Canada, but mainly in eastern USA, and extending down to South America. It is probably the most severe encephalitis virus in humans and horses. EEE is principally an infection of birds, and in North America is spread among them mainly by *Culiseta melanura* (the genus *Culiseta* is not discussed in this book). It is transmitted to people and horses by various species of *Ochlerotatus, Culex* and *Coquillettidia*.

Western equine encephalitis (WEE) (*Alphavirus*). In western USA and extending into South America. This is basically an arboviral infection of birds, which are *amplifying* and *maintenance* hosts. Transmission is by *Cx. tarsalis*, a ricefield breeding mosquito, as well as by other *Culex* and *Ochlerotatus* species. *Ochlerotatus* mosquitoes also transmit the virus to mammals, including humans and horses.

Venezuelan equine encephalitis (VEE) (*Alphavirus*). In southern USA through Central America to northern parts of South America. The infection is often fatal in horses but usually very mild in humans. Rodents are believed to be important *amplifying hosts* involved in enzootic transmission cycles, and birds and bats may be involved in dispersing the virus. Vectors are mainly *Culex* species. Epizootic/epidemic transmission involves equines which are the principal amplifying hosts. *Ochlerotatus taeniorhynchus* and *Psorophora confinnis* appear to be the main vectors in epizootics.

Viraemic titres produced in humans by JE, EEE, SLE, VEE and sometimes also by WEE are so low that the infection cannot be transmitted by mosquitoes from humans to humans or from humans to other susceptible hosts. Thus humans are often referred to as *'dead-end' hosts*. Similarly horses are 'dead-end' hosts for the encephalitis viruses infecting them.

3.8.6   Filariasis

The development in mosquitoes of filarial worms causing lymphatic filariasis is briefly described in Chapter 2 in connection with the role of anopheline vectors. Both bancroftian and brugian filariasis occur in two distinct forms. The *nocturnal periodic* form, in which the microfilariae are in the peripheral blood only at night, is transmitted by night-biting mosquitoes such as *Anopheles*, *Mansonia* and *Culex quinquefasciatus*. During the day the microfilariae are in the blood vessels supplying the lungs, and are not available to be taken up by mosquitoes. In *subperiodic* forms of *Wuchereria bancrofti* and *Brugia malayi* the microfilariae exhibit a reduced periodicity and are present in the peripheral blood during the day as well as at night, but there nevertheless remains a degree of periodicity. For example, subperiodic *W. bancrofti*, such as found in Polynesia, has a small peak in microfilarial density during the daytime and can therefore be called *diurnal subperiodic*, whereas subperiodic *B. malayi* in West Malaysia, Sumatra, Sabah, Thailand, etc. has a slight peak of microfilariae at night, and so can be called *nocturnal subperiodic*.

The list of vectors of lymphatic filariasis is complex, and some 40 *Anopheles* species and 40 culicine species belonging to five genera have been incriminated as vectors of bancroftian and brugian filariasis. Only some of the more important species can be given in the following account and in Table 3.1. For a more detailed account of the vectors see the table in Zagaria and Savioli (2002).

Bancroftian filariasis

Wuchereria bancrofti occurs throughout much of the tropics (Haiti, Dominican Republic, coastal areas of Guyana and Brazil, Africa, Asia, Pacific areas)

Table 3.1. *Summary of principal mosquito vectors of filariasis*

Species and forms of filariasis	Geographical distribution	Vectors	Zoonotic reservoirs
Wuchereria bancrofti			
Nocturnal periodic	Throughout tropics (except Polynesia)	*Anopheles* species *Culex quinquefasciatus*	None
	Papua New Guinea	As above and *Mansonia uniformis*	
	Philippines	*Ochlerotatus poicilius* and *Anopheles* species	
	China	*Ochlerotatus togoi*[a]	
Diurnal subperiodic	Polynesia in general	*Aedes polynesiensis*	None
	Fiji	*Aedes pseudoscutellaris*	
	New Caledonia	*Ochlerotatus vigilax*	
Nocturnal subperiodic	Thailand	*Ochlerotatus niveus*	None
Brugia malayi			
Nocturnal periodic (principally open swamps)	South Asia, from India eastwards	*Anopheles* species, *Mansonia annulata, Ma. annulifera, Ma. uniformis*	Not important, possibly some exist
	China	*Ochlerotatus togoi*[a]	
Nocturnal subperiodic (mainly in swampy forests)	Malaysia, Indonesia, Thailand and Philippines	*Mansonia annulata, Ma. bonneae, Ma. dives, Coquillettidia crassipes*	Monkeys, especially leaf-monkeys (*Presbytis* species), wild and domestic cats, and pangolins
Brugia timori			
Nocturnal periodic	Timor, Flores and other Indonesian islands	*Anopheles barbirostris*	None

[a] *Ochlerotatus togoi* is now a very minor vector because the incidence of filariasis in China has been greatly reduced.

and also in the Middle East, mainly Egypt. It is the most widely distributed filarial infection of humans. Bancroftian filariasis is essentially an urban disease; there are no animal reservoirs and the parasites develop only in mosquitoes and humans.

The **nocturnal periodic** form is transmitted by *Anopheles* species (Chapter 2) and throughout much of its distribution also by *Culex quinquefasciatus*. This mosquito is widespread in the tropics and breeds mainly in waters polluted with human or animal faeces or rotting vegetation and other filth; larvae are found in septic tanks, cesspits, pit latrines, drains and ditches, and in water storage jars if they contain organically polluted water. It is a mosquito that has increased in numbers in many towns due to rapid and increasing urbanization and the resultant proliferation of insanitary collections of water. Adults bite at night and after feeding they often rest in houses. Although *Cx. quinquefasciatus* is a good vector in most areas of Africa, it is a poor vector in West Africa where most transmission of bancroftian filariasis is by anopheline mosquitoes.

In Papua New Guinea *Mansonia uniformis* and night-biting *Culex* species are also vectors.

In the Philippines *Ochlerotatus poicilius* is the most important vector. Adults bite in the early part of the night, mainly indoors but also sometimes outdoors. After feeding, adults rest outdoors. Larvae occur in leaf axils of banana, plantain and coco-yam (*Colocasia*) plants. In China filariasis has been greatly reduced, and so the former vector *Ochlerotatus togoi* is now just a minor vector.

The **diurnal subperiodic** form occurs in the Polynesian region, from where the nocturnal periodic form is absent. The most important vector is *Aedes polynesiensis*, a day-biting mosquito which feeds mostly outdoors but may enter houses to feed; adults rest almost exclusively out of doors. Larvae occur in natural containers such as split coconut shells, leaf bracts and crab-holes, and also in man-made containers such as discarded tins, pots, vehicle tyres and canoes. *Aedes pseudoscutellaris* is another outdoor day-biting mosquito that is a vector of diurnal subperiodic *W. bancrofti*, especially in Fiji. It breeds mainly in tree-holes and bamboo stumps but larvae are also found in crab-holes. In New Caledonia *Ae. polynesiensis* is absent and the most important vector is *Ochlerotatus vigilax*, adults of which feed outdoors mainly during the day. Larvae are found in brackish or fresh water in rock-pools and ground pools.

The **nocturnal subperiodic** form is found in Thailand and is transmitted by the *Ochlerotatus niveus* group of mosquitoes. Adults bite and rest outdoors; larvae are found mainly in bamboo.

It should be noted that although several *Aedes* mosquitoes are vectors of filariasis, especially the bancroftian form, *Ae. aegypti* is not a vector of lymphatic filariasis. Infection rates of mosquitoes with infective larvae of *W. bancrofti* range usually from about 0.1% to 5%, depending greatly on vector species and local conditions. But in Singapore infection rates of 20% have been recorded.

Brugian filariasis

The **nocturnal periodic** form is principally a rural disease. It occurs throughout most of Asia, from southern India to Malaysia, Vietnam, Cambodia, Thailand, Indonesia and southern China. It is transmitted mainly by night-biting *Mansonia* mosquitoes, such as *Ma. annulata*, *Ma. annulifera* and *Ma. uniformis*, and also by various *Anopheles* species (Chapter 2). These *Mansonia* species breed in more-or-less permanent waters such as swamps and ponds having floating or rooted aquatic vegetation. Adults bite mainly outdoors and rest out of doors after feeding, but they will also bite and rest indoors in some areas. *Ochlerotatus togoi* was a vector of nocturnal periodic *B. malayi* in China but is now a minor vector because of the reduction of filariasis. There are no known important animal reservoirs of the nocturnal periodic form.

The **nocturnal subperiodic** form of *B. malayi* occurs in west Malaysia, Indonesia, Thailand and the Philippines, and is transmitted by *Mansonia* mosquitoes, mainly by *Ma. annulata*, *Ma. bonneae* and *Ma. dives*, and in Thailand and the Philippines also by *Coquillettidia crassipes*. Larvae of all these species occur in habitats with much vegetation, such as swampy forests. Adults bite mainly at night but also during the daytime, especially species such as *Ma. dives* and *Ma. bonneae*. This subperiodic form of *B. malayi* is essentially a parasite of swamp monkeys, especially the so-called leaf monkeys (*Presbytis* species), humans becoming infected when they live at the edges of these areas. Other reservoirs include *Macaca* monkeys, domestic and wild cats, such as civets, and pangolins (i.e. scaly anteaters, genus *Manis*).

Infection rates of mosquitoes with infective larvae of *B. malayi* range from about 0.1% to 2–3%, which is slightly lower than for *W. bancrofti*, but infection rates vary according to mosquitoes and local conditions.

As mentioned in Chapter 2, the discovery of filarial worms in mosquitoes does not necessarily imply they are vectors of either *W. bancrofti* or *B. malayi* because mosquitoes are also vectors of several other filarial parasites of animals.

3.9   CONTROL

Repellents, mosquito nets and mosquito screening of houses and other personal protection measures (discussed in Chapters 1 and 2) can give some relief from culicine mosquitoes. It is, however, more difficult to obtain protection from culicines than from anophelines because many of them bite outdoors during the daytime. Spraying the interior surfaces of houses with residual insecticides, as practised for *Anopheles* control, is not usually applicable to culicines, because most species do not rest predominantly in houses. The main method of attack involves larval control, although some-times aerial ultra-low-volume (ULV) applications are used to kill adult culicine mosquitoes.

3.9.1   *Aedes*, *Ochlerotatus* and *Psorophora*

Control of mosquitoes such as *Ae. aegypti*, *Ae. polynesiensis* and *Ae. albopic-tus*, species which breed mainly in man-made containers in both rural and urban areas, is often aimed at reducing the numbers of larval habitats, that is control by **source reduction**. Thus, people are encouraged not to store water in pots or allow water to accumulate in discarded tin cans, bottles, vehicle tyres, etc. However, persuading people to cooperate in reducing peridomestic breeding of mosquitoes is often unsuccessful unless local legislation is strictly enforced. A reliable piped water supply to houses can do much to reduce *Ae. aegypti* breeding.

When source reduction is not feasible insecticides can be used. Although insecticidal spraying will kill *Aedes*, *Ochlerotatus* and *Psorophora* larvae it usually has no effect on their dry, but viable, eggs. These have been deposited at the edges of larval habitats and will hatch when the water level rises and floods them. *Aedes*, *Ochlerotatus* and *Psorophora* mosquitoes breed-ing in pools, ponds and marshy areas can be controlled by ground-based or aerial applications of granular organophosphate or carbamate insecti-cides, or insect growth regulators. Applications can be made either before or after habitats have become flooded, that is **pre- or post-flood treatments**. Insecticidal granules landing on dry or muddy grounds remain more or less inactive until the habitats become flooded. When this occurs previ-ously dry eggs hatch, but at the same time flooding results in the release of insecticide from the granules and this kills the newly hatched larvae. This technique helps overcome problems of controlling *Aedes*, *Ochlerotatus* and *Psorophora* mosquitoes whose eggs may hatch in instalments over extended periods after flooding.

Ground-based or aerial ULV applications of malathion, fenitrothion, pirimiphos-methyl, propoxur or synthetic pyrethroids is often the most appropriate control strategy in epidemic situations.

When insecticides are used to kill mosquito larvae in water that is intended for drinking, it is essential that only those having extremely low mammalian toxicity are used; they should also impart no taste to the water. The insecticide usually recommended is temephos, in the form of 1% in briquettes or sand granules that will give a concentration of 1 mg active ingredient per litre of water. However, some communities refuse to have their potable water dosed with any insecticide.

Yellow fever and dengue

Africa has over 90% of the world's yellow fever cases, and of about 60 epidemics recorded in the past 70 years, 60% were in the 1990s. The best defence against yellow fever is to use the attenuated 17D vaccine. Vector control, through sustained reduction in mosquito breeding, nevertheless still has a role in reducing the risks of yellow fever outbreaks; it also remains the main approach for dengue control, because there are no vaccines ready for field use. Epidemics of dengue, and sometimes also yellow fever, can be curtailed by killing the adult vectors by aerial ULV insecticidal spraying, where the main objective is to kill as rapidly as possible infected adult mosquitoes.

3.9.2 *Culex*

Culex quinquefasciatus, an important filariasis vector, is best controlled by improving sanitation and installing modern sewage systems, but often this is not feasible and insecticidal measures have to be employed. In most areas *Cx. quinquefasciatus* is resistant to a wide range of insecticides and this limits the choices of chemicals that can be used. Larval habitats should be sprayed every 7–10 days, and usually relatively large dosage rates are needed because most insecticides are less effective in the presence of organic pollution, which is characteristic of *Cx. quinquefasciatus* breeding places. Chlorpyrifos is usually one of the more effective insecticides in polluted waters. Synthetic pyrethroids and insect growth regulators such as methoprene and diflubenzuron have also been used against *Cx. quinquefasciatus*.

Tipping non-toxic expanded ***polystyrene beads*** (2–3 mm) into pit latrines and cesspits to completely cover the water surface with a 2–3 cm layer suffocates larvae and pupae as well as preventing female *Cx. quinquefasciatus* from ovipositing. A single application can persist for several years and give excellent control. This is a control method that is readily accepted by most communities.

Insecticidal house-spraying, as practised against malaria, and insecticide-impregnated bed-nets can be effective against *Cx. quinquefasciatus* and other *Culex* species if they are both endophilic and night-biters.

In North America ULV spraying has frequently been used against vectors of the encephalitis viruses.

Lymphatic filariasis

In 1997 the WHO passed a resolution calling for the eradication of lymphatic filariasis and in the year 2000 the plan was endorsed and adopted by the Global Programme to Eliminate Lymphatic Filariasis (GPELF). The date set to achieve this goal is the year 2020.

Control strategies involve mass distribution of microfilarial drugs (albendazole with either ivermectin or DEC) given annually or in cooking salt, and concurrently reducing vector populations. Application of suitable larvicides, including the microbial insecticides *Bacillus thuringiensis* var. *israelensis* and *B. sphaericus*, use of expanded polystyrene beads in pit latrines and septic tanks, use of insecticide-impregnated bed-nets, and possibly residual house-spraying, can all reduce densities of *Cx. quinque-fasciatus*. In contrast it is very difficult to effectively reduce populations of *Mansonia* vectors, because of their often large and relatively inaccessible larval habitats and their larval biology, and because adults are not generally endophilic.

3.9.3 *Mansonia*

Mansonia mosquitoes are usually controlled by removing, or using herbicides to kill, aquatic weeds upon which the larvae and pupae depend for their oxygen requirements. Altering aquatic habitats by such methods may, however, result in ecological changes that allow them to become colonized by mosquito species that were previously excluded by the dense covering of weeds.

If insecticides are used to control *Mansonia* larvae, then granules or pellets are more suitable than liquid formulations. This is because they can penetrate the vegetation, sink to the bottom of breeding places and release their chemicals through the water. However, species such as *Ma. dives* and *Ma. bonneae*, which are important vectors of brugian filariasis, breed in extensive swampy forests, and are usually impossible to control because of the large and often inaccessible areas involved.

FURTHER READING

Anon. (1998) Main theme: dengue. *Public Health*, **14**, 6–58. Germany: Bayer A. G.

Chan, K. L. (1985) *Singapore's Dengue Haemorrhagic Fever Programme: a Case Study on the Successful Control of* Aedes aegypti *and* Aedes albopictus *Using Mainly Environmental Measures as a Part of Integrated Vector Control*. Publication no. 5. Tokyo: Southeast Asian Medical Information Center.

Curtis, C. F., Malecela-Lazaro, M., Reuben, R. and Maxwell, C. A. (2002) Use of floating layers of polystyrene beads to control populations of the filaria vector *Culex quinquefasciatus*. *Annals of Tropical Medicine and Parasitology*, **96** (suppl. 2), 97–104.

Gratz, N. and Knudsen, A. B. (1996) *The Rise and Spread of Dengue, Dengue Haemorrhagic Fever and its Vectors: a Historical Review (up to 1995)*. CTD/FIL(DEN) 96.7. Geneva: World Health Organization.

Gubler, D. J. and Kuno, G. (eds.) (1997) *Dengue and Dengue Haemorrhagic Fever*. Wallingford: CAB International.

Halstead, S. B. and Gomez-Dantes, H. (eds.) (1992) *Dengue: a World-Wide Problem, a Common Strategy*. Proceedings of the international conference on dengue and *Aedes aegypti* community-based control. Mexico: Ministry of Health, Rockefeller Foundation.

Monath, T. P. (ed.) (1988) *Arboviruses: Epidemiology and Ecology. Volume 1: General Principles. Volume 2: African Horse Sickness to Dengue. Volume 3: Eastern Equine Encephalomyelitis to O'nyong nyong. Volume 4: Oropouche*

Fever to Venezuelan Equine Encephalomyelitis. Volume 5: Vesicular Stomatitis to Yellow Fever. Boca Raton, FL: CRC Press.

Monath, T. P. (2001) Yellow fever. In: *The Encyclopedia of Arthropod-transmitted Infections of Man and Domesticated Animals*, ed. M. W. Service. Wallingford: CABI. pp. 571–7.

Muller, R. (2002) *Worms and Human Diseases*. 2nd edn. Wallingford: CABI.

Reeves, W. C. (ed.) (1990) *Epidemiology and Control of Mosquito-Borne Arboviruses in CA, 1943–1987*. Sacramento, CA: California Mosquito and Vector Control Association.

Tomori, O. (1999) Impact of yellow fever on the developing world. *Advances in Virus Research*, **53**, 5–34.

White, G. B. and Nathan, M. B (eds.) (2002) The elimination of lymphatic filariasis: public-health challenges and the role of vector control. *Annals of Tropical Medicine and Parasitology*, **96** (suppl. 2), 1–164.

World Health Organization (1997) *Dengue Haemorrhagic Fever: Diagnosis, Treatment, Prevention and Control*. 2nd edn. Geneva: World Health Organization.

Zagaria, N. and Savioli, L. (2002) Elimination of lymphatic filariasis: a public-health challenge. *Annals of Tropical Medicine and Parasitology*, **96** (suppl. 2), 3–13.

See also references at the ends of Chapters 1 and 2.

4

Black-flies (Simuliidae)

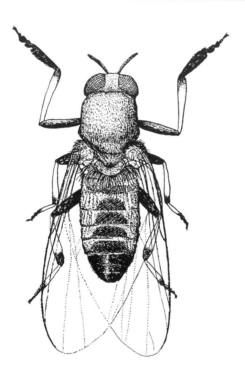

Black-flies belong to the family Simuliidae. They have a worldwide distribution. There are almost 1800 species in about 25 genera. However, only
four genera – *Simulium, Prosimulium, Austrosimulium* and *Cnephia* – contain
species that bite people.

Medically, *Simulium* is by far the most important genus as it contains
many vectors. In Africa, species in the *S. damnosum* complex and *S. neavei*
group, and in Central and South America, species in the *S. ochraceum,*
S. metallicum and *S. exiguum* complexes, transmit the parasitic nematode
Onchocerca volvulus which causes human onchocerciasis (river blindness).
In Brazil, *S. amazonicum* transmits *Mansonella ozzardi*, a filarial parasite that
is usually regarded as non-pathogenic.

4.1   EXTERNAL MORPHOLOGY

The Simuliidae are commonly known as black-flies, but in some areas, especially Australia, they may be called sand-flies. As explained in Chapter 5
this latter terminology is confusing and best avoided because biting flies
in the family Ceratopogonidae (biting midges) are also sometimes called
sand-flies, while flies in the subfamily Phlebotominae are regarded as the
true sand-flies.

Adult black-flies are quite small, about 1.5–4 mm long, relatively stout-
bodied and, when viewed from the side, have a rather *humped* thorax. As
their vernacular name indicates they are usually *black* in colour but some
species have contrasting patterns of white, silvery or yellowish hairs on
their bodies and legs, while others may be predominantly orange or bright
yellow.

Black-flies have a pair of large compound eyes, which in females are
separated on the top of the head (a condition known as *dichoptic*); in the
males the eyes occupy almost all of the head, and touch on top of it and
in front above the bases of the antennae (a condition known as *holoptic*).
In males, but not females, the lenses are larger on the upper than on the
lower half of the eyes (Fig. 4.1b). The antennae are short, stout, cylindrical
and distinctly segmented (usually 11 segments) but lack long hairs. The
mouthparts are short and relatively inconspicuous but the five-segmented
maxillary palps hang downwards and are easily seen. Only females bite.
The morphology of the mouthparts is similar to those of the biting midges
(Ceratopogonidae, Chapter 6). The mouthparts are short and broad and
do not penetrate very deeply into the host's tissues. Teeth on the labrum
stretch the skin, while the rasp-like action of the maxillae and mandibles
cuts through the skin and ruptures its fine blood capillaries. The small pool
of blood produced is then sucked up by the flies. This method of feeding is
ideally suited for picking up the microfilariae of *Onchocerca volvulus*, which
occur in human skin, not blood.

The thorax is covered dorsally with very fine and appressed hairs, which
can be black, white, silvery, yellow or orange and may be arranged in

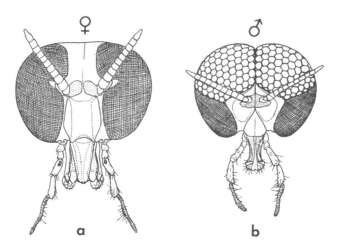

Figure 4.1 Front view of adult *Simulium* heads; (a) female with dichoptic eyes; (b) male with holoptic eyes.

various patterns. The relatively short legs are also covered with very fine and closely appressed hairs and may be unicolorous or have contrasting pale and dark bands. Each tarsus has a pair of claws, which are untoothed (i.e. simple) in mammal-feeders. The wings are characteristically short and broad and *lack* both scales and prominent hairs. Only the veins near the anterior margin are well developed; the rest of the wing is membranous and has indistinct venation (Fig. 4.2). The wings are colourless or almost so, and when at rest are closed over the body like the blades of a closed pair of scissors.

The abdomen is short and squat, and covered with inconspicuous closely appressed fine hairs. In neither sex are the genitalia very conspicuous. Black-flies are most easily *sexed* by looking at their eyes.

4.2   LIFE CYCLE

Eggs are about 0.1–0.4 mm long, brown or black, and are more or less triangular in shape but have rounded corners and smooth unsculptured shells (Fig. 4.3a) and are covered with a sticky substance. They are always laid in *flowing water* but the type of breeding place differs greatly according to species. Habitats can vary from small trickles of water, slow-flowing streams, lake outlets and water flowing from dams to fast-flowing rivers and rapids. Some species prefer lowland streams and rivers whereas others are found in mountain rivers. In species such as *S. ochraceum*, a Central American vector of onchocerciasis, eggs are scattered over the surface of flowing water while females are in flight. In most species, however, ovipositing females alight on partially immersed objects such as rocks, stones and vegetation to lay some 150–800 eggs in sticky masses or strings. Females may crawl underneath the water and become completely

Figure 4.2 Adult simuliid black-fly (*Simulium damnosum*) in lateral view (courtesy of R. W. Crosskey and the Natural History Museum, London).

submerged during oviposition. There may be a few favoured oviposition sites in a stream or river, resulting in thousands of eggs from many females being found together. *Simulium damnosum*, for example, frequently has such communal oviposition sites. Eggs are unable to survive desiccation.

Eggs of *S. damnosum* hatch within about 1–2 days but in many other tropical species the egg stage lasts 2–4 days. Eggs of species inhabiting temperate or cold northern areas may not hatch for many weeks and some species pass the winter as diapausing eggs.

There are six to nine (usually seven) larval instars and the mature larva, depending on the species, is about 4–12 mm long. It is easily distinguished from all other aquatic larvae (Fig. 4.3b). The head is black, or almost so, and has a prominent pair of *feeding brushes* (cephalic fans), while the weakly segmented, cylindrical body is usually greyish, but may be darker or sometimes even greenish. The body is slightly swollen beyond the head and in most, but not all, species distinctly swollen towards its end. The rectum has finger-like rectal organs which on larval preservation may be extruded and visible as a dorsal protuberance towards the end of the abdomen. Ventrally,

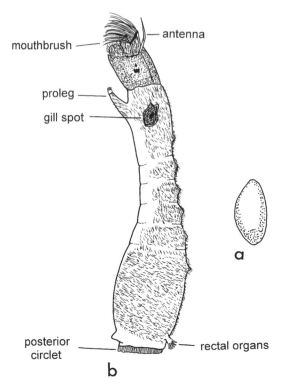

Figure 4.3 (a) *Simulium* egg; (b) lateral view of the last larval instar showing principal diagnostic simuliid characters. The species figured is *Simulium damnosum*, a species which has the body covered with minute dark setae and has small dorsal tubercles, i.e. small humps.

just below the head, is a small thoracic *proleg* which is armed with a small circlet of hooklets.

Larvae do not swim but remain sedentary for long periods on submerged vegetation, rocks, stones and other debris. Attachment is achieved by the *posterior hook-circlet* (anal sucker of many previous authors) tightly gripping a small silken pad which has been produced by the larva's very large salivary glands and which is firmly glued to the substrate. Larvae can nevertheless move about. This is achieved by alternately attaching themselves to the substrate by the proleg and the posterior hook-circlet, which results in their moving in a looping manner. When larvae are disturbed they can deposit sticky saliva on a submerged object, release their hold and be swept downstream for some distance at the end of a silken thread. They can then either swallow the thread of saliva and regain their original position, or reattach themselves at a site further downstream.

Larvae normally orientate themselves to lie *parallel* to the flow of water with their heads downstream. They are mainly filter-feeders, ingesting,

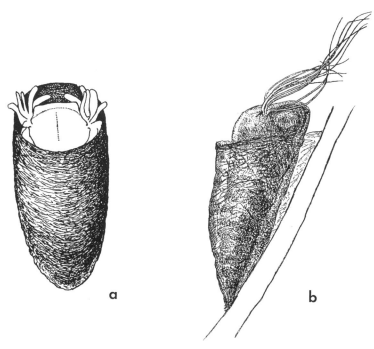

Figure 4.4 *Simulium* pupae in their cocoons: (a) dorsal view of a species (*S. damnosum*) with broad and short respiratory filaments (courtesy of R. W. Crosskey and the Natural History Museum, London); (b) lateral view of a species with long and thin respiratory filaments.

with the aid of large feeding brushes, suspended particles of food. However, a few species have predacious larvae and others are occasionally cannibalistic. Depending on species and temperature larval development may be as short as 6–12 days, but in some species it may be extended to several months, and in other species larvae overwinter.

Mature larvae, which can be recognized by a blackish mark termed a *gill spot* (respiratory organ of the future pupa) on each side of the thorax (Fig. 4.3b), spin, with the silk produced by the salivary glands, a protective slipper-shaped brownish *cocoon*. The cocoon is firmly attached to submerged vegetation, rocks or other objects and its shape and structure vary greatly according to species (Fig. 4.4). After weaving the cocoon the enclosed larva pupates. The pupa has a pair of thin-walled *respiratory gills* which are usually prominent and may be filamentous or broad. Their length, shape and the number of filaments or branches provide useful taxonomic characters for species identification. These gills, and the anterior part of the pupa, often project from the entrance of the cocoon (Fig. 4.4). In both tropical and non-tropical countries the pupal period lasts only 2–6 days and is unusual in not appearing to be dependent on temperature.

On emergence adults either rise rapidly to the water surface in a protective bubble of gas, which prevents them from being wetted, or they escape by crawling up partially submerged objects such as vegetation or rocks. A characteristic of many species is the almost simultaneous mass emergence of thousands of adults. On reaching the water surface the adults immediately take flight.

The empty pupal cases, with gill filaments still attached, remain enclosed in their cocoons after the adults have emerged and retain their taxonomic value, by providing means of identifying simuliid species that have successfully emerged.

A few African and Asian black-fly species have a very unusual aquatic association. For example, in East Africa larvae and pupae of *S. neavei* do not occur on submerged rocks or vegetation but on other aquatic arthropods, such as the bodies of immature stages (nymphs) of mayflies (Ephemeroptera), and various crustaceans including fresh-water crabs. Such an association is termed a **phoretic** relationship. As eggs are never found on these animals they are probably laid on submerged stones or vegetation (p. 87).

The nuclei of the larval salivary gland cells have large **polytene chromosomes** which have banding patterns that are used to identify otherwise morphologically identical species within a species complex. For example, chromosomal studies have shown that there are about 40 cytologically different entities in the *S. damnosum* complex, many of which are distinct species, whereas other forms are best referred to as cytoforms until their reproductive isolation, or otherwise, is determined.

4.2.1   Adult behaviour

Both male and female black-flies feed on plant juices and naturally occurring sugary substances, but only females take blood-meals. Biting occurs out of doors during the day. Many species, including *S. damnosum* in Africa, have bimodal biting patterns, with a peak in the early morning and another in the afternoon or early evening. However, in some species, such as *S. ochraceum* in Guatemala, biting continues more or less throughout the day. Many species seem particularly active on cloudy, overcast days and in thundery weather. Black-flies may exhibit marked preferences for feeding on different parts of the body; for example, *S. damnosum* feeds mainly on the legs whereas *S. ochraceum* prefers to bite the head and torso. When feeding on animals, adults crawl down the fur of mammals, or feathers of birds, to bite the host's skin; they may also enter the ears to feed.

Many species of black-fly feed almost exclusively on birds (**ornithophagic**), others on non-human mammalian hosts (**zoophagic**), while several species also bite people (**anthropophagic**). Some of these, however, actually prefer various large animals such as donkeys or cattle and bite humans only as a poor second choice, whereas others appear to find humans almost equally attractive hosts; no species feeds exclusively on people. In many species vision seems important in host location but host odours also play a role. After feeding, blood-engorged females shelter in vegetation, on trees and in other natural outdoor resting places until the blood-meal is completely digested. In the tropics this takes 2–3 days, while in non-tropical areas it may take 3–8 days or longer, the speed of digestion depending mainly on temperature. A few species can lay eggs without a blood-meal (i.e. they are *autogenous*). Relatively little is known about black-fly longevity, but it seems that adults of most species live for 3–4 weeks.

Females of some species may fly considerable *distances* (15–30 km) from their emergence sites to obtain blood-meals and may also be dispersed long distances by winds. For example, it is not exceptional for adults of *S. damnosum* to be found biting 60–100 km from their breeding places, and in West Africa there is evidence that prevailing winds can carry adults up to 400–600 km. The long distances involved in dispersal have great relevance in control programmes, because areas freed from black-flies can be reinvaded from distant breeding places. In Central and South America black-flies generally disperse only about 2–15 km.

In temperate and northern areas of the Palaearctic and Nearctic regions biting nuisance from simuliids is seasonal, because adults die in the autumn and new generations do not appear until the following spring or early summer. Although in many tropical areas there is continuous breeding throughout the year, there may nevertheless be dramatic increases in population size during the rainy season.

4.3   MEDICAL IMPORTANCE

4.3.1   Annoyance

In both tropical and non-tropical areas of the world black-flies can cause a very serious biting problem. Although the severity of the reaction to bites differs in different individuals, localized swelling and inflammation frequently occurs, accompanied by intense irritation lasting for several days or even weeks. Repeated biting by black-flies such as *S. erythrocephalum* in central Europe, *S. posticatum* in England, *S. venustum* and *S. vittatum* in North America can cause headaches, fevers, swollen lymph glands and aching joints. In some areas of North America outdoor activities are almost impossible at certain times of the year due to the intolerable numbers of biting simuliids. The classical example of the nuisance that can result from black-flies was the seasonal exodus during the eighteenth century of people from the Danube valley area in central Europe, largely to save their livestock from attack by the enormous numbers of *S. colombaschense*. However, it is as disease vectors that black-flies are most important medically.

4.3.2   Onchocerciasis

Onchocerciasis is a non-fatal disease, often called river blindness, that is caused by the filarial parasite *Onchocerca volvulus*. There are no animal reservoirs, so the disease is not a zoonosis. Onchocerciasis occurs throughout West Africa, Central Africa, and much of East Africa from Ethiopia to Tanzania with isolated pockets of infection in Malawi, Sudan and southern Yemen, possibly extending into Saudi Arabia. About 95% of all cases occur in 27 countries in Africa. In the Americas onchocerciasis is localized in areas of southern Mexico, Guatemala, Brazil, Venezuela, Ecuador and Colombia.

Black-flies are the only vectors of human onchocerciasis. Their habit of tearing and rasping the skin to rupture blood capillaries when obtaining a blood-meal makes them particularly suited to the ingestion of the **skin-borne microfilariae** of *O. volvulus*. Most of the microfilariae ingested during feeding are destroyed or excreted, but some penetrate the stomach wall and migrate to the thoracic muscles. Here they develop into sausage-shaped stages (called L_1 larvae), then moult to L_2 stage larvae, and a few of these moult again to become elongate and thinner L_3 worms which pass through the head and down the proboscis. These *infective third-stage* worms (about 660 μm in length) leave the proboscis and penetrate the host's skin when black-flies feed on a person. The interval between the ingestion of microfilariae to the time infective larvae (L_3) are in the proboscis is about 6–12 days, depending on temperature.

African vectors of onchocerciasis

Chromosomal studies show that the *S. damnosum* complex is composed of about 40 cytoforms, many of which merit species status. The *damnosum* complex is widespread in tropical Africa and certain of its species are the most important vectors of onchocerciasis.

Adults of the *damnosum* complex are mainly black. They can be recognized by the front tarsi being broad and flattened, having a conspicuous dorsal crest of fine hairs, and by the presence of a very broad white area on the first segment of the hind tarsus (basitarsus) (Fig. 4.2). Larvae are more or less covered with fine setae and there are usually prominent dorsal abdominal tubercles (Fig. 4.3b). Branches of the pupal respiratory gills are thick and finger-like and are contained within the neck of the cocoon (Fig. 4.4a). The immature stages are found in the rapids of small or very large rivers in both savannah and forested areas of Africa. Adults frequently disperse hundreds of kilometres from their breeding sites. The vectorial status of species within the *damnosum* complex varies, the most important vectors of onchocerciasis being *S. damnosum* s. str., *S. sirbanum*, *S. sanctipauli* and *S. leonense*. Other, but much less important, vectors include *S. neavei* which is responsible for transmission in the Congo, Zaire and Uganda, and formerly in Kenya, where it was eradicated in the 1950s by bush clearing and insecticidal dosing of streams.

Simulium neavei is a **phoretic** species, that is the larvae and pupae are found attached to other freshwater fauna, in this instance crabs of the genus *Potamonautes* and mayfly (Ephemeroptera) nymphs, both of which occur in small rocky, rather turbid streams and rivers.

American vectors of onchocerciasis

Simulium ochraceum is widely distributed in Central America and northern parts of South America and is the principal onchocerciasis vector in southern Mexico and Guatemala. Adults are very small and are easily recognized by their dark brown legs, bright orange scutum (dorsal surface of the thorax) and yellow basal part of the abdomen, which contrasts with the black apical part. Females oviposit while in flight, dropping their eggs onto floating vegetation. Larval habitats consist of trickles of flowing water and very small streams which are often concealed by bushes, vegetation and fallen leaves. Adults do not appear to disperse far. The main biting season is unusual in being in the drier months of the year.

Simulium metallicum occurs in Mexico through Central America to northern areas of South America. In Venezuela it appears to be the most important vector whereas in other areas such as in Mexico and Guatemala it is considered a minor vector. It is a black species and has a broad white area on the first segment of the hind tarsus. Larvae occur in small or large streams and rivers. Adults fly further from their breeding sites than do those of *S. ochraceum*.

Simulium exiguum is the only known vector in Colombia, and a primary vector in Ecuador and a secondary one in Venezuela. Above 150 m in the Brazilian Amazonas the main vector is *S. guianense*; below this height the vector is *S. oyapockense*.

4.3.3   *Mansonella ozzardi*

Mansonella ozzardi is a filarial parasite of humans that is usually regarded as non-pathogenic, although it has been reported as causing morbidity in Colombia and Brazil. It is transmitted in the Caribbean islands, Trinidad, Surinam and Argentina by *Culicoides* species, mainly *C. furens* and *C. phlebotomus* (Chapter 6), but in Brazil, Colombia, Guyana, Venezuela and southern Panama *S. amazonicum* is the main vector.

4.4   CONTROL

Some protection can be gained by using repellents such as DEET, or by wearing pyrethroid-impregnated clothing.

Although insecticidal fogging of vegetation harbouring resting adult black-flies has occasionally been undertaken, this approach results in very temporary and localized control. The only practical control method is the application of insecticides to larval breeding places. Insecticides need be applied to only a few selected sites on watercourses for some 15–30 minutes, because as the insecticide is carried downstream it kills simuliid larvae over long stretches of water. Flow rates of the water and its depth are used to calculate the quantity of insecticide to be released. In the past dosing rivers with DDT has given good control of *S. damnosum* in Africa and contributed to the eradication of *S. neavei* in Kenya. However, because of its accumulation in food chains DDT is no longer recommended for larval control; instead organophosphates, such as temephos or *Bacillus thuringiensis* var. *israelensis* (*B. thuringiensis* serotype H-14) are usually recommended. Treatment has to be regularly repeated, sometimes at 1–2 week intervals, throughout the year to prevent recolonization.

In many areas ground application of larvicides is difficult, either because of the enormous size of the rivers requiring treatment or because breeding occurs in a large network of inaccessible small streams and watercourses. Under these conditions aerial applications from small aircraft or helicopters may be appropriate

4.4.1   Onchocerciasis control programme (OCP)

Because of the severity of river blindness in the Volta River Basin area of West Africa and its devastating effect on rural life the world's most ambitious and largest vector control programme, the Onchocerciasis Control Programme (OCP), was initiated in 1974 by the World Health Organization. By 1986 there were 11 participating countries (Benin, Burkina Faso, Côte d'Ivoire, Ghana, Guinea, Guinea Bissau, Mali, Niger, Senegal, Sierra Leone and Togo). Some 50 000 km of rivers over an area of 1.3 million km^2 that were breeding the *S. damnosum* complex were dosed weekly with temephos, which was dropped from helicopters or small aircraft. Because of the appearance of temephos resistance in 1980 in some populations and species of the *S. damnosum* complex, some rivers were treated with other insecticides or with *Bacillus thuringiensis* var. *israelensis*. To hinder the spread of further resistance a rotation of different insecticides was initiated in 1982. Larviciding continued in the countries added in 1986 (Guinea, Guinea Bissau, Senegal, Sierra Leone) until the year 2002. Since 1988 the OCP has been undertaking large-scale distribution of the microfilaricidal drug *ivermectin* (Mectizan), which is given orally once or twice a year. Results from the OCP have been most spectacular, and transmission of river blindness has ceased over most of the OCP area.

In 1995 the African Programme for Onchocerciasis Control (APOC) was created to cover populations at risk in 19 countries outside the OCP. The objective is to establish, within 12 years, a sustainable community-based ivermectin treatment regimen, backed up with focal larviciding. Because ivermectin does not kill adult worms control needs to continue for about

20 years to allow time for the reservoir of infection (adult onchocercal worms) in the human population to die out.

FURTHER READING

Anon. (1998) Mectizan and onchocerciasis: a decade of accomplishment and prospects for the future. The evolution of a drug into a development concept. *Annals of Tropical Medicine and Parasitology*, **92** (suppl. 1), 1–179.

Crosskey, R. W. (1990) *The Natural History of Blackflies*. Chichester: Wiley.

Davies, J. B. (1994) Sixty years of onchocerciasis vector control: a chronological summary with comments on eradication, reinvasion, and insecticide resistance. *Annual Review of Entomology*, **39**, 23–45.

Davies, J. B. and Crosskey, R. W. (1991) *Simulium* – vectors of onchocerciasis. Mimeographed document WHO/VBC/91.992. Geneva: World Health Organization.

De Villiers, P. C. (1987) *Simulium* dermatitis in man: clinical and biological features in South Africa. *South African Medical Journal*, **71**, 523–5.

Hougard, J.-M., Yaméogo, L., Sékétéli, A., Boatin, B. and Dadzie, K. Y. (1997) Twenty-two years of black-fly control in the onchocerciasis control programme in West Africa. *Parasitology Today*, **13**, 425–8.

Molyneux, D. H. and Davies, J. B. (1997) Onchocerciasis control: moving towards the Millennium. *Parasitology Today*, **13**, 418–25.

Raybould, J. N. and White, G. B. (1979) The distribution, bionomics and control of onchocerciasis vectors (Diptera: Simuliidae) in eastern Africa and the Yemen. *Tropenmedizin und Parasitologie*, **30**, 505–47.

Service, M. W. (1977) Methods for sampling adult Simuliidae, with special reference to the *Simulium damnosum* complex. *Tropical Pest Bulletin*, **5**, 1–48.

World Health Organization (1995) Onchocerciasis and its control. Report of a WHO expert committee on onchocerciasis control. *World Health Organization Technical Report Series*, **852**, 1–103.

(1997) *Twenty Years of Onchocerciasis Control: Review of the Work of the Onchocerciasis Control Programme in West Africa from 1974 to 1994*. Geneva: World Health Organization. (In English or French.)

5

Phlebotomine sand-flies (Phlebotominae)

There are some 600 species of phlebotomine sand-flies in six genera within the subfamily Phlebotominae of the family Psychodidae. Species in three genera – *Phlebotomus*, *Lutzomyia* and *Sergentomyia* – suck blood from vertebrates; the former two are the more important as they contain disease vectors. The genus *Phlebotomus* occurs only in the Old World, from southern parts of the northern temperate areas, such as the Mediterranean region, to tropical areas, but there are not many species in sub-Saharan Africa, especially West Africa. Most *Phlebotomus* species inhabit semiarid and savannah areas in preference to forests. *Lutzomyia* species by contrast are found only in the New World tropics, occurring mostly in the forested areas of Central and South America.

Sergentomyia species are also confined to the Old World, being especially common in the Indian subregion, but also occurring in other areas such as Africa and central Asia. A few species of *Sergentomyia* bite people, but they are not disease vectors.

Adult flies are often called sand-flies. However, this can be confusing because in some parts of the world the small biting midges of the Ceratopogonidae (Chapter 6) and black-flies (Simuliidae, Chapter 4) are called sand-flies. The medically most important species include *Phlebotomus papatasi*, *P. sergenti*, *P. argentipes*, *P. ariasi*, *P. perniciosus* and species in the *Lutzomyia longipalpis* and *Lutzomyia flaviscutellata* complexes. In both the Old and New Worlds sand-flies are vectors of leishmaniasis and viruses responsible for sandfly fever, and in the Andes the bacterium *Bartonella bacilliformis,* causing bartonellosis (Carrión's disease).

5.1   EXTERNAL MORPHOLOGY

No details are given for distinguishing between the adults of *Phlebotomus* and *Lutzomyia* because this requires specialized knowledge and detailed examination; generally species biting people in the Old World will be *Phlebotomus* and in the New World *Lutzomyia*.

Adult phlebotomine sand-flies can be readily recognized by their minute size (1.3–3.5 mm in length), hairy appearance, relatively large black eyes and their long and *stilt-like* legs. The only other blood-sucking flies which are as small as this are some species of biting midges (Ceratopogonidae), but these have non-hairy wings and differ in many other details (Chapter 6). Phlebotomine sand-flies have the head, thorax, wings and abdomen densely covered with long hairs. The antennae are long and composed of small bead-like segments having short hairs; antennae are similar in both sexes. The mouthparts are short and inconspicuous and adapted for blood-sucking, but only females bite. At their base is a pair of five-segmented maxillary palps which are relatively conspicuous and droop downwards.

Wings are lanceolate in outline and quite distinct from the wings of other biting flies. The Phlebotominae can be distinguished from other subfamilies of the family Psychodidae, which they may superficially resemble, by the

Figure 5.1 Adult male phlebotomine sand-fly showing genital claspers at end of abdomen, and a diagrammatic representation of the double branching of wing vein 2.

wings. In sand-flies the wings are held erect over the body when the fly is at rest, whereas in non-biting psychodid flies they are folded, roof-like, over the body. The wing venation also differs. In phlebotomine sand-flies, but not in the other subfamilies of Psychodidae, vein 2 *branches twice*, although this may not be apparent unless most of the hairs are rubbed from the wing veins (Fig. 5.1).

The abdomen is moderately long and in the female more or less rounded at the tip. In males it terminates in a prominent pair of genital claspers which give the end of the abdomen an upturned appearance.

Identification of adult phlebotomine sand-flies to species is difficult and usually necessitates the examination of internal structures, such as the arrangement of the teeth on the cibarial armature, the shape of the spermatheca in females, and in males the structure of the external genitalia (terminalia).

5.2   LIFE CYCLE

The minute eggs (0.3–0.4 mm) are more or less ovoid in shape and usually brown or black, and careful examination under a microscope reveals that they are patterned, as shown in Figure 5.2. Some 30–70 eggs are laid singly at each oviposition. They are deposited in small cracks and holes in the ground, at the base of termite mounds, in cracks in masonry, on stable floors, in poultry houses, amongst leaf litter and in between buttress roots of forest trees, etc. The type of oviposition site varies greatly according to species.

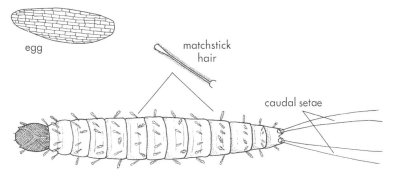

Figure 5.2 Egg of a phlebotomine sand-fly showing mosaic-type pattern, and a larva with matchstick hairs and caudal setae.

Although eggs are not laid in water they require a moist microhabitat with a high humidity. They are unable to withstand desiccation and hatch after 4–20 days, although hatching may be prolonged in cooler weather. Larvae are mainly scavengers, feeding on organic matter such as fungi, decaying forest leaves, semi-rotting vegetation, animal faeces and decomposing bodies of arthropods. Although some species, especially of the genus *Phlebotomus*, occur in semiarid areas, the actual larval habitats must have a high degree of humidity. Larvae can usually survive if their breeding places are temporarily flooded by migrating to drier areas.

There are four larval instars. The mature larva is 3–6 mm long and has a well-defined black head which is provided with a pair of small mandibles; the body is white or greyish and 12-segmented (Fig. 5.2). Ventrally the abdominal segments have small pseudopods, but the most striking feature of the larva is the presence on the head and all body segments of conspicuous thick bristles with feathered stems, which in many species have slightly enlarged tips. They are called **matchstick** hairs and identify the larvae as those of phlebotomine sand-flies. In most species the last abdominal segment bears two pairs of conspicuous long hairs called the *caudal setae*. The first-instar larva has two single bristles not two pairs.

Larval development is usually complete after 20–30 days, the duration depending on species, temperature and availability of food. In temperate areas and arid regions species may overwinter as diapausing fully grown larvae. Prior to pupation the larva assumes an almost erect position in the habitat, the skin then splits open and the pupa wriggles out. The larval skin is not completely cast off but remains attached to the end of the pupa. The presence of this skin, with its characteristic two pairs of caudal bristles, aids in the recognition of the phlebotomine pupae. The pupal shape is as shown in Figure 5.3. Adults emerge from the pupae after about 6–13 days. The life cycle, from oviposition to adult emergence, lasts 30–60 days, but extends to several months in some species with diapausing larvae. For

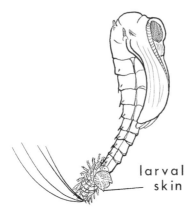

larval
skin

Figure 5.3 Pupa of a phlebotomine sand-fly with larval skin still attached.

instance, in temperate areas adults die off in late summer or autumn and species overwinter as larvae, with the adults emerging the following spring. Apart from this the duration of the life cycle may be longer during cooler periods of the year.

It is usually extremely difficult to find larvae or pupae of sand-flies and relatively little is known about their biology and ecology.

5.2.1   Adult behaviour

Both sexes feed on plant juices and sugary secretions but *females* in addition suck blood from a variety of vertebrates, including livestock, dogs, urban and wild rodents, snakes, lizards and amphibians; a few species feed on birds. In the Old World females of many *Phlebotomus* species, and in the tropical Americas *Lutzomyia* species, bite mammals, including humans. Biting is usually restricted to crepuscular and nocturnal periods but people may be bitten during the day in darkened rooms, or in forests during overcast days. Most species feed out of doors (*exophagic*) but some species also feed indoors (*endophagic*). A few species are autogenous, that is they can lay eggs without blood-feeding.

Adults are *weak fliers* and do not usually disperse more than a few hundred metres from their breeding places. Consequently biting is often very localized. However, adults of at least some species can fly up to 2.2 km over a few days. Sand-flies have a characteristic hopping type of flight so that there may be several short flights and landings before females settle on their hosts. Windy weather inhibits flight activities and biting. Because of their very short mouthparts sand-flies are unable to bite through clothing.

During the day adult sand-flies rest in sheltered, dark and humid sites, but on dry surfaces, such as on tree trunks, on ground litter and foliage of forests, in animal burrows, termite mounds, tree-holes, rock fissures, caves, cracks in the ground and inside human and animal habitations. Species that commonly rest in houses (*endophilic*) before or after feeding on humans are often referred to as domestic or peridomestic species. Examples are *Phlebotomus papatasi* in the Mediterranean area and the *Lutzomyia longipalpis* complex in South America.

In temperate areas of the Old World sand-flies are seasonal in their appearance and adults occur only in the summer months. In tropical areas some species are common more or less throughout the year, but in other species there may be well-marked changes in abundance of adults related to the dry and wet seasons.

5.3   MEDICAL IMPORTANCE

5.3.1   Annoyance

Apart from their importance as disease vectors, sand-flies may constitute a serious, but usually localized, biting nuisance. In previously sensitized people their bites may result in severe and almost intolerable irritations, a condition known in the Middle East as harara.

5.3.2   Leishmaniasis

Leishmaniasis is a term used to describe a number of closely related diseases caused by about 20 distinct species, subspecies and strains of *Leishmania* parasites in more than 100 countries. The three main clinical forms are cutaneous, mucocutaneous and visceral leishmaniasis. A fourth, less common form is termed diffuse cutaneous leishmaniasis, while post-kala-azar dermal leishmaniasis is caused by *Leishmania donovani donovani* following cure of the initial visceral form. The epidemiology of leishmaniasis is complex, involving not only different parasite species but different strains of parasites and different reservoir hosts.

Basically **amastigote** parasites ingested by female sand-flies with a blood-meal multiply in the gut and develop into **promastigotes**. These are elongate, have a flagellum and attach to the mid-gut or hind-gut wall and multiply rapidly. Many, however, are voided when the fly defecates. After further development the survivors migrate to the anterior part of the mid-gut and from there to the fore-gut. Here some parasites become **metacyclic** forms. Four to twelve days after the sand-fly has taken an infective blood-meal the metacyclic forms are found in the mouthparts, from where they are introduced into a new host during feeding. Infective flies often probe more often than uninfected flies, thus maximizing transmission of parasites during blood-feeding. Previous feeding by females on sugary substances, mostly obtained from plants, is essential not only for the survival of the sand-fly but for the development of the parasites to the infective form.

Most types of leishmaniasis are **zoonoses**. The degree of involvement of humans varies greatly from area to area. The epidemiology is largely determined by the species of sand-flies, their ecology and behaviour, the availability of a wide range of non-human hosts and also by the species and strains of *Leishmania* parasites. In some areas, for example, sand-flies will transmit infections almost entirely among wild or domesticated animals, with little or no human involvement, whereas elsewhere animals may be important *reservoir hosts* of infection for humans. In India the disease may be transmitted between people by sand-flies, with animals taking no identifiable part in its transmission. The epidemiology of the leishmaniases is complex, and only simplified accounts are given below.

Cutaneous leishmaniasis (CL). In the Old World, CL is known also as oriental sore. Occurs mainly in arid areas of the Middle East to north-western India and central Asia, in North Africa and various areas in East, West and southern Africa. The principal parasites are *Leishmania major*, transmitted mainly by *Phlebotomus papatasi*, and *Le. tropica*, transmitted by *P. sergenti*. *Leishmania major* is usually zoonotic and in most of its range gerbils are the reservoir hosts; *Le. tropica* occurs in densely populated areas and humans appear to be the main reservoir hosts. In the New World, CL is found mainly in forests from Mexico to northern Argentina, and is caused by *Leishmania braziliensis*, *Le. amazonensis* and *Le. mexicana*. Rodents and dogs appear to be reservoir hosts. Vectors include *Lutzomyia wellcomei* and *L. flaviscutellata*.

Mucocutaneous leishmaniasis (ML) (espundia). A severely disfiguring disease found from Mexico to Argentina. It is mainly caused by *Leishmania braziliensis*. Dogs may be reservoir hosts. *Lutzomyia wellcomei* is an important vector.

Diffuse cutaneous leishmaniasis (DCL). A form that causes widespread cutaneous nodules or macules over the body. It is confined to Venezuela and the Dominican Republic and the highlands of Ethiopia and Kenya. In South America the parasite is *Le. amazonensis* transmitted by *Lutzomyia flaviscutellata*; spiny rats are reservoir hosts. In Ethiopia and Kenya the parasite is *Le. aethiopica*, transmitted by *Phlebotomus pedifer* and *P. longipes,* with rock hyraxes as reservoir hosts.

Visceral leishmaniasis (VL). Often referred to as kala-azar. It is caused by *Leishmania donovani donovani* in most areas of its distribution, such as India, Bangladesh, Sudan, East Africa and Ethiopia. Among the vectors are *P. argentipes* and *P. orientalis*. Rodents and wild cats may be reservoir hosts. In the Mediterranean basin, Iran and central Asia, including northern and central China, *Leishmania donovani infantum* (*Le. chagasi* of some authors) is the parasite, and the vectors include *P. ariasi* and *P. perniciosus*. Dogs and foxes are reservoir hosts. Visceral leishmaniasis also occurs sporadically in Central and South America, where the parasite is *Leishmania donovani infantum*, transmitted by *Lutzomyia longipalpis*. Foxes may be reservoir hosts.

5.3.3   Bartonellosis

Bartonellosis, sometimes called Oroya fever or Carrión's disease, is encountered in arid mountainous areas of the Andes, mainly in Peru, but also in Ecuador and Colombia. It is caused by the bacterium *Bartonella bacilliformis* and is transmitted by *Lutzomyia verrucarum* and possibly by other *Lutzomyia* species. Transmission is probably only by contamination of the mouthparts.

5.3.4   Sandfly fevers

Sand-flies transmit the seven viral serotypes responsible for sandfly fevers, also called papataci fever (sometimes spelt papatasi or pappataci fever), three-day fever or *Phlebotomus* fevers. The classical form of the disease is found in the Mediterranean region, but also extends up the Nile into Egypt, and from the Middle East to northern India, Pakistan, Afghanistan and probably China. The most important vector is *P. papatasi*. Other forms of the virus in Central and South America are transmitted by *Lutzomyia* species such as *L. trapidoi.*

Female sand-flies become infective 7–10 days after taking a blood-meal. Infected females can lay eggs containing the virus and these eventually give rise to infected adults. This is an example of **transovarial** transmission, a phenomenon that is more common in the transmission of various tick-borne diseases (Chapter 17). There are probably mammalian reservoir hosts.

5.4   CONTROL

There have been very few organized attempts to control transmission of leishmaniasis. Phlebotomine sand-flies are, however, very susceptible to insecticides. In nearly all areas where residual insecticides have been used to control *Anopheles* vectors of malaria there have been dramatic reductions in sand-fly populations followed by interruption of leishmaniasis transmission. With the cessation of spraying, sand-flies usually return, and transmission is renewed. Obviously where sand-flies bite and rest out of doors house-spraying will have little effect.

If the outdoor resting sites are known (e.g. animal shelters, stone walls, tree trunks, termite hills) they can be sprayed with residual insecticides. Insecticidal fogging of outdoor resting sites may give some, but temporary, control of vectors.

Personal protection can be achieved by applying efficient insect repellents such as DEET. Mesh screens or nets with very small holes can give protection, but they reduce ventilation causing it to be unpleasantly hot in screened houses or under sand-fly bed-nets. Such nets, or even mosquito nets with larger holes, can be impregnated with pyrethroids (e.g. permethrin and deltamethrin) to give protection against biting for up to 6 months (see p. 48 for more details on the use of insecticide-impregnated bed-nets for protection against malaria vectors).

Control of sand-fly larvae remains impossible because the breeding sites of most species are unknown, and even when they can be identified it is usually impractical to obtain effective vector control.

Because most leishmaniasis transmission involves *reservoir* animals, such as rodents and dogs, attempts have sometimes been made to destroy such hosts. In China leishmaniasis was effectively eliminated in the 1950s by killing dogs, but although similar culls have been made in parts of Brazil and the Mediterranean region results have been disappointing. In Russia and Jordan zoonotic cutaneous leishmaniasis was controlled by destroying rodent colonies, but elsewhere results have not been encouraging.

FURTHER READING

Ashford, R. W. (2001) Leishaniasis. In: *The Encyclopedia of Arthropod-Transmitted Infections of Man and Domesticated Animals*, ed. M. W. Service. Wallingford: CABI. pp. 269–79.

Hide, G., Mottram, J. C., Coombs, G. H. and Holmes, P. H. (1996) *Trypanosomiasis and Leishmaniasis: Biology and Control*. Wallingford: CAB International.

Killick-Kendrick, R. (1990) Phlebotomine vectors of the leishmaniases: a review. *Medical and Veterinary Entomolology*, **4**, 1–24.

(1999) The biology of phlebotomine sand flies. *Clinics in Dermatology*, **17**, 279–89.

Lainson, R. (1983) The American leishmaniases: some observations on their ecology and epidemiology. *Transactions of the Royal Society of Tropical Medicine and Hygiene*, **77**, 569–96.

(1989) Demographic changes and their influence on the epidemiology of the American leishmaniases. In: *Demography and Vector-Borne Diseases*, ed. M. W. Service. Boca Raton, FL: CRC Press. pp. 85–106.

Lane, R. P. (1991) The contribution of sand-fly control to leishmaniasis control. *Annales de la Société Belge de Médicine Tropicale*, **71** (suppl.), 65–74.

Peters, W. and Killick-Kendrick, R. (eds) (1987) *The Leishmaniases in Biology and Medicine. Volume 1: Biology and Epidemiology. Volume 2: Clinical Aspects and Control*. London: Academic Press.

Ward, R. D. (1990) Some aspects of the biology of phlebotomine sand-fly vectors. *Advances in Disease Vector Research*, **6**, 91–126. (Reprints and chapters incorrectly dated 1989.)

World Health Organization (1990) Control of the leishmaniases: report of a WHO Expert Committee. *World Health Organization Technical Report Series*, **793**, 1–158.

6

Biting midges (Ceratopogonidae)

Biting midges belong to the family Ceratopogonidae, of which there are about 5500 species in some 125 genera. Species in only four genera feed on vertebrates. Medically the most important are the genera *Leptoconops*, which is mainly found in the tropics and subtropics, including the Caribbean area and parts of the USA, and *Culicoides* which contains over 1400 species and has an almost worldwide distribution. In many parts of the world species of *Culicoides*, and in the Americas also *Leptoconops*, can constitute serious biting problems. In Africa species of the *Culicoides milnei* group (e.g. *C. milnei* and *C. austeni*), *C. grahamii* and *C. inornatipennis* are vectors of the filarial worms *Mansonella perstans*, while *C. grahamii* and possibly the *C. milnei* group are vectors of *Mansonella streptocerca*. *Culicoides furens* is a vector of *Mansonella ozzardi* in the Americas. These parasites are usually regarded as non-pathogenic to humans. Although they can be a serious biting nuisance *Leptoconops* are not disease vectors.

6.1   EXTERNAL MORPHOLOGY

A generalized description is given of the adults of *Culicoides*. Adult *Culicoides* are sometimes known as midges or biting midges, and, especially in the Americas, as 'no-see-ums' or 'punkies', and in Australia and some other countries as sand-flies. This last name is unfortunate and should be avoided because phlebotomines (Chapter 5) and occasionally simuliids (Chapter 4) may also be referred to as sand-flies. The most appropriate common name is biting midges; this terminology serves to distinguish them from other small non-biting flies which are often referred to as midges.

The adults are very small insects only 1–2.5 mm long, and with the phlebotomines constitute the smallest biting flies attacking humans.

The head bears a prominent pair of eyes, a pair of short five-segmented palps and a pair of relatively long filamentous antennae. As in mosquitoes, males do not take blood-meals and have feathery or *plumose antennae*, whereas the blood-sucking females have *non-plumose* antennae. The biting mouthparts are very small and inconspicuous and they do not project forwards but hang down vertically from the head. The arrangement and structure of the mouthparts are very similar to those of simuliids (Chapter 4). In some species the thorax is covered dorsally with very small but distinct black spots and markings. In addition to these dark markings, a pair of shiny black small but elongated depressions, known as *humeral pits*, are present in all *Culicoides* species on the dorsal surface of the anterior part of the thorax. The presence of these pits distinguishes the genus *Culicoides* from *Leptoconops*. The wings are short and relatively broad, and apart from the first veins their venation is faint. Wings lack scales but in many species are covered with minute hairs – these are seen only under a dissecting microscope – and in most *Culicoides* species the density of these hairs gives the *wings* the appearance of having contrasting dark and milky white spots or patches (Fig. 6.1). In life the wings are placed over the abdomen

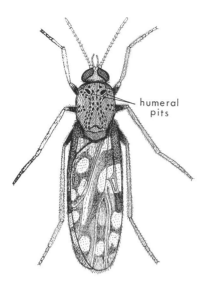

humeral
pits

Figure 6.1 Adult *Culicoides* midge showing patterned wings and the two thoracic humeral pits.

like the blades of a closed pair of scissors (Fig. 6.1). The legs are relatively short.

The abdomen is dull grey, yellowish brown or blackish and in the female is more or less rounded at its tip; in the male there is a pair of small, but conspicuous, claspers.

6.2   LIFE CYCLE

Eggs are dark, cylindrical or curved and banana-shaped, and about 250–500 μm long (Fig. 6.2c). They are laid in batches of about 30–350 on the surface of mud, wet soil, especially that near swamps and marshes including saltwater marshes, on decaying leaf litter, humus, manure, or on plants and other objects near or partially submerged in water, and also in treeholes, in semi-rotting vegetation and in the cut stumps of banana plants (for example *Culicoides milnei* group and *C. grahamii*). The type of oviposition site selected depends on the species.

Eggs usually hatch after 2–9 days, depending on temperature and species; some temperate species overwinter as eggs. There are four larval instars and a fully grown larva is cylindrical, whitish, about 3–6 mm long and nematode-like. It has a small, dark (but unpigmented in *Leptoconops*) conical-shaped head followed by 12 body segments. The last segment terminates in a pair of four-lobed *retractile papillae* (Fig. 6.2a). These are not always readily seen in preserved larvae because they are often retracted within the last abdominal segment (Fig. 6.2b). *Culicoides* larvae are best recognized by the combination of a small dark head followed by a segmented

Figure 6.2 *Culicoides*: (a) larva with two four-lobed retractile papillae extending beyond the last abdominal segment; (b) last abdominal segment with papillae withdrawn into it; (c) eggs.

body devoid of any obvious structures and, when they are extruded, by the presence of terminal papillae. When alive they can also be recognized by their serpentine swimming motions.

Larvae feed mainly on decaying vegetable matter and occur in many different types of habitats, including freshwater or saltwater marshes and swamps, edges of ponds, boggy and semi-waterlogged areas, and in specialized habitats such as horse and cow excreta, tree-holes and rotting cacti. When the water level in swamps and marshes rises the larvae of many species migrate to the damp soil and mud at the edges to avoid becoming completely submerged. Some important pest species breed in sandy areas near the seashore. Larvae are difficult to find and are rarely encountered unless special surveys are made to collect them.

Larval development is sometimes completed within 4–5 days, but may take weeks in cooler weather, and in temperate regions many species overwinter and remain as larvae for 7 months or more. Species occupying marshy habitats frequently migrate to the drier peripheral areas for pupation. However, in species that are aquatic the pupae float at the water surface.

The pupa (Fig. 6.3) is 2–4 mm long and readily recognized by the following combination of characters: (1) a pair of breathing trumpets on the cephalothorax which appear to be composed of two segments; (2) abdominal segments bearing small but conspicuous tubercles ending in a fine hair; and (3) a prominent pair of horn-like processes on the last abdominal segment. The pupal period usually lasts 2–3 days.

6.2.1   Adult behaviour

Adults of both sexes feed on naturally occurring sugar solutions. In addition females take blood-meals from humans and a variety of mammals and birds. Adults bite at any time of the day or night, but many species are particularly active and troublesome in the evenings and first half of the night. In contrast *C. grahamii* bites mainly in the early part of the mornings. Because of their short mouthparts, biting midges are not as successful

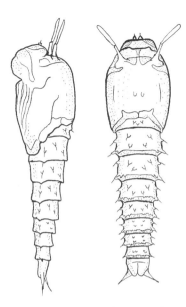

Figure 6.3 *Culicoides* pupa, lateral and dorsal views. Note the two long respiratory trumpets.

in biting through clothing as are mosquitoes, tabanids and tsetse-flies, all of which have a longer proboscis. For this reason midges often appear in swarms or clouds around the head, biting the face, especially the forehead and scalp. They also bite other exposed parts such as the hands and arms, and the legs of people wearing shorts. Most species bite only outdoors, but a few including the *C. milnei* group and *C. grahamii* will enter houses to feed on people (*endophagic*). About 30 species are *autogenous*, that is they can mature and lay the first batch of eggs without a blood-meal. Adults normally fly only a few hundred metres from their larval habitats, but sometimes they get dispersed considerably further by the wind. However, some species of both *Culicoides* and *Leptoconops* are known to fly 2–3 km without the assistance of wind. Adult females are short-lived, probably surviving for just 1–3 weeks.

6.3　MEDICAL IMPORTANCE

The Ceratopogonidae are not considered very important vectors of disease to humans, whereas they are of considerable veterinary importance, trans-mitting arboviruses such as those causing bluetongue disease in sheep and African horse sickness. They can, however, be very troublesome pests and transmit minor filarial infections as well as Oropouche virus to humans (see below).

6.3.1   Annoyance

Biting midges are very small, but what they lack in size they can make up for in numbers – as has been said, one midge is an entomological curiosity, a thousand sheer hell! In several areas of the world, as dissimilar as the west coast of Scotland, the Caribbean and the sunny regions of California and Florida, biting midges can be a serious economic threat to the tourist industry. The persistent biting of large numbers of midges can make out-door recreational activities impossible, not only at dusk but often during much of the day. In some areas they have even prevented the continuation of harvesting and other outdoor work during the evenings.

Important pest species in southern areas of North America down to Brazil include *C. furens* which breeds in salt marshes and other saline coastal habitats. In North America *Leptoconops torrens* and *L. bequaerti*, which also breed in sandy soils and coastal areas, can be serious pests. In Europe *C. impunctatus* and many other species are troublesome biters, while in Madagascar, the Seychelles and Brunei *L. spinosifrons* can cause a considerable biting problem.

6.3.2   Filarial infections

A few *Culicoides* species are vectors of filarial parasites to humans.

> ***Mansonella perstans***. Found in Africa, especially West and Central, but also parts of East Africa as far south as Zimbabwe. Vectors are species in the *C. milnei* group, including *C. milnei* and *C. austeni*, and possibly also *C. grahamii*. *Mansonella perstans* also occurs in Trinidad and South America where it is transmitted by other *Culicoides* species.
>
> ***Mansonella streptocerca***. Occurs in the rain forests of West Africa, from Côte d'Ivoire to Gabon, Congo and Zaire and also in western Uganda, where transmission is by *C. grahamii*, and possibly by the *C. milnei* group, such as *C. milnei* and *C. austeni*. These species breed in the rotting cut stumps of banana and plantain plants.
>
> ***Mansonella ozzardi***. Found in Mexico, Panama, the Caribbean islands and South America. Transmission is by *Culicoides* species, mainly *C. furens* and *C. phlebotomus*. (The role of simuliids in the transmission of this filarial parasite is discussed in Chapter 4.)

Microfilariae of all these parasites are ***non-periodic***, they are ingested with a blood-meal and pass through a similar developmental cycle to other filarial parasites in mosquitoes. That is, they undergo morphological changes, invade the thoracic flight muscles, moult twice and then migrate to the head and after about 8–12 days pass down the proboscis. The infective ***third-stage*** larvae (L$_3$) are deposited on the skin of the host when the female takes a blood-meal. The salivary glands of *Culicoides* play no part in the transmission of these parasites. None of the three filarial parasites carried by midges appears to cause much harm and they are usually regarded as non-pathogenic, although morbidity or allergic reactions may sometimes occur.

6.3.3   Arboviruses
The only known arbovirus biologically transmitted to humans by midges is Oropouche virus (*Bunyavirus*) which occurs in Brazil, Panama, Peru, Trinidad and possibly Colombia. Transmission is by the bite of *Culicoides paraensis.*

6.4   CONTROL
Because many of the major pest species breed in extensive and often diffuse habitats, such as freshwater and saltwater marshes and wet coastal sands, whose limits may be difficult to define, it is usually very difficult to reduce larval breeding substantially.

Although larval habitats can be eradicated by draining them or filling them in, this is usually costly, laborious and in many areas impractical. Sometimes semi-aquatic sites such as muddy or marshy areas can be impounded and flooded under 5–8 cm of water to ensure that the soil is never exposed, thus destroying suitable habitats. This type of environmental control can be effective against some species but ineffective against others; it can also be very expensive. If maintained, such methods have the advantage of giving permanent control. However, although they avoid contaminating the environment with insecticides they in themselves result in completely changing the habitat.

Insecticidal spraying of larval breeding sites has sometimes been effective, but heavy rainfall is required to wash the insecticide through the surface vegetation to the underlying soil and mud harbouring the midge larvae. Increased public awareness of environmental contamination would these days likely prevent such indiscriminate insecticidal applications.

Insecticidal fogs or ultra-low-volume (ULV) applications have sometimes been used to kill adults resting in vegetation, but the effects are very short-lived and sprayed areas are soon invaded by midges flying in from unsprayed areas.

Limited personal protection can be achieved by the use of repellents such as DEET. A few species enter buildings to bite but mosquito nets and screening used to keep out house-flies and mosquitoes will not prevent biting by the much smaller midges, unless they are impregnated with pyrethroids such as permethrin and deltamethrin. Such treated nets may remain effective for 6 months or more, after which they need to be re-impregnated. Alternatively protective nets and screens having very small mesh size can be used, but they will substantially reduce air flow and ventilation.

FURTHER READING

Holbrook, F. R. (1985) An overview of *Culicoides* control. In: *Bluetongue and Related Orbiviruses*, ed. T. L. Barber and M. M. Jochim. Progress in Clinical and Biological Research 178. New York, NY: Liss. pp. 607–9.

Kettle, D. S. (1965) Biting ceratopogonids as vectors of human and animal diseases. *Acta Tropica*, **22**, 356–62.

(1969) The ecology and control of blood-sucking ceratopogonids. *Acta Tropica*, **26**, 235–48.

(1977) Biology and bionomics of blood-sucking ceratopogonids. *Annual Review of Entomology*, **22**, 33–51.

Linley, J. R. and Davies, J. B. (1971) Sandflies and tourism in Florida and the Bahamas and Caribbean area. *Journal of Economic Entomology*, **64**, 264–78.

Linley, J. R., Hoch, A. L. and Pinheiro, F. P. (1983) Biting midges (Diptera: Ceratopogonidae) and human health. *Journal of Medical Entomology*, **20**, 347–64.

Mellor, P. S. (2001) Oropouche virus. In: *The Encyclopedia of Arthropod-Transmitted Infections of Man and Domesticated Animals*, ed. M. W. Service. Wallingford: CABI. pp. 391–3.

Mellor, P. S., Boorman, J. and Baylis, M. (2000) *Culicoides* biting midges: their role as arbovirus vectors. *Annual Review of Entomology*, **45**, 307–40.

Nathan, M. B. (1979) The prevalence and distribution of *Mansonella ozzardi* in coastal north Trinidad, W. I. *Transactions of the Royal Society of Tropical Medicine and Hygiene*, **73**, 299–302.

Reynolds, D. G. and Vidot, A. (1978) Chemical control of *Leptoconops spinosifrons* in the Seychelles. *Pest Articles and News Summaries*, **24**, 19–26.

7

Horse-flies (Tabanidae)

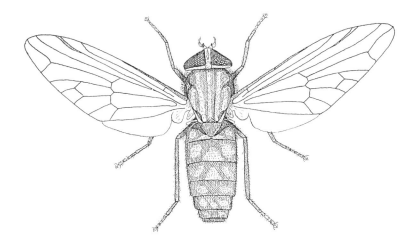

Tabanids are large biting flies generally called horse-flies, although other vernacular names include greenheads (some species of *Tabanus*), clegs and stouts (*Haematopota*) and deer-flies (*Chrysops*). They all belong to the family Tabanidae, which comprises some 4300 species in about 130 genera. The most important species medically belong to the genera *Tabanus*, *Chrysops* and *Haematopota*. Tabanids have been incriminated in the spread of anthrax and in North America and Russia also tularaemia, and it has been suggested they might be involved in the transmission of Lyme disease (usually transmitted by hard ticks). But the main medical importance of the Tabanidae is that species of *Chrysops*, mainly *C. silaceus* and *C. dimidiatus*, are vectors in West and Central Africa of the filarial worm *Loa loa*.

The Tabanidae have a worldwide distribution. Species of *Tabanus* and *Chrysops* are found in temperate and tropical areas, but *Haematopota* is absent from South America and Australasia and is uncommon in North America.

7.1   EXTERNAL MORPHOLOGY

A generalized description is presented of the Tabanidae, with special reference to the genera *Chrysops*, *Tabanus* and *Haematopota*.

Tabanids are medium to very large flies (6–30 mm long). Many, especially of the genus *Tabanus*, are robust and heavily built, and this genus contains the largest biting flies, some attaining a wingspan of 65 mm. The colouration of tabanids varies from very dark brown or black to lighter reddish-brown, yellow or greenish; frequently the abdomen and thorax have stripes or patches of contrasting colours (Fig. 7.1). The head is large and, viewed from above, is more or less *semicircular* in outline (Fig. 7.2); it is often described as semilunar. The head bears a conspicuous pair of compound eyes which in life may be marked with contrasting *iridescent* colours, such as greens and reds or even purplish hues, arranged in bands, zig-zags or spots. Adults are sexed by examination of their eyes. In the female there is a distinct space on top of the head between the eyes: this is known as a *dichoptic* condition (Fig. 7.2a). In females of some species this space between the eyes may be narrow, whereas in others, especially *Chrysops*, it is quite large. In males the eyes are so large that they occupy almost all of the head and either touch each other on top of the head or are very narrowly separated, this being known as a *holoptic* condition (Fig. 7.2b).

The *antennae* are relatively small but stout. They consist of three segments; the last is subdivided into usually three or four small divisions by annulations. Unlike the Muscidae, Glossinidae and Calliphoridae there is no antennal arista. The size and shape of the antennae serve to distinguish the genera *Chrysops*, *Haematopota* and *Tabanus* (Fig. 7.3). The mouthparts of female Tabanidae are stout and adapted for biting and, unlike those

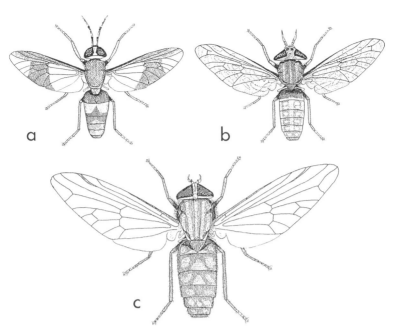

Figure 7.1 Adult female tabanids belonging to three genera: (a) *Chrysops*; (b), *Haematopota*; (c) *Tabanus*.

Figure 7.2 Dorsal view of tabanid heads; (a) female showing dichoptic eyes: (b) male with holoptic eyes.

of tsetse-flies, mosquitoes and *Stomoxys*, they do not project forwards but hang downwards from the head. Only female tabanids take blood-meals.

The thorax is stout and bears a pair of wings which have two sub-marginal and five posterior cells and a completely closed *discal cell* in approximately the centre of the wing (Fig. 7.4). Although the wing venation alone may not be sufficient to identify the Tabanidae from all other Diptera, it nevertheless serves as a useful guide when considered with other characters, such as the shape and structure of the antennae and biting mouthparts. The wings may be completely clear and devoid of colour or have areas of brown colouration, or they may be distinctly banded,

Figure 7.3 Antennae of adult tabanids belonging to three genera: (a) *Chrysops*; (b) *Haematopota*; (c) *Tabanus*.

Figure 7.4 Wing of an adult tabanid showing discal cell, two sub-marginal cells (subm.) and five posterior cells (1–5).

or appear mottled or speckled due to the presence of greyish patches (Fig. 7.1). When adults are at rest the wings are placed either like a pair of open scissors over the abdomen or at a roof-like angle completely obscuring the abdomen. The presence or absence of coloured areas on the wings and the way in which they are held over the body provides useful additional characters for distinguishing between *Chrysops*, *Tabanus* and *Haematopota* (see below).

The abdomen is usually broad and stout, and in unfed flies characteristically flattened dorsoventrally. It may be a more or less uniformly dark brown, blackish, light brown, reddish brown, yellowish or even greenish, or alternatively marked with contrasting coloured stripes or patches.

7.1.1   Identification of adult *Chrysops*, *Tabanus* and *Haematopota*

Chrysops species (deer-flies)

Chrysops species are medium-sized flies ranging in size from that of a house-fly to a tsetse-fly. In life most species have *iridescent* eyes, commonly with spots of red, green or purple. The wings, which are held partially over the abdomen in an open scissor-like fashion, usually have one or more brownish transverse bands (Fig. 7.1a). In many species the abdomen is blackish with orange or yellow patches or bands.

The most reliable method of distinguishing *Chrysops* species from *Tabanus* and *Haematopota* is by the *antennae*. In *Chrysops* these are long, the second segment is long and cylindrical and has no projection (Fig. 7.3a), and the third segment has four small subdivisions. The hind tibiae have apical spurs; these are absent in *Tabanus* and *Haematopota*.

Chrysops has a worldwide distribution.

Tabanus species (horse-flies, greenheads)

Tabanus species are medium to very large flies. The *eyes* are frequently brownish but may be iridescent; the markings are usually in the form of horizontal bands. The wings, which are held over the body much as in *Chrysops*, are often clear (Fig. 7.1c), but in some species there are dark markings.

Tabanus species are readily identified by the shape and size of the **antennae**. Both the second and third antennal segments have small but distinct projections on the upper surface (Fig. 7.3c), and the third segment has four small subdivisions and is usually distinctly curved upwards. The antennae are much shorter than those of *Chrysops* species, and are therefore less conspicuous.

Tabanus has a worldwide distribution.

Haematopota species (clegs, stouts)

Haematopota species are medium-sized dark grey flies which are easily distinguished from *Tabanus* and *Chrysops* by the fact that in life the wings are folded *roof-like* over the abdomen. Moreover, in nearly all species the wings are **dusty grey** and speckled or mottled (Fig. 7.1b). The *eyes* have zig-zag bands of iridescent colours. Antennae are rather similar to those of *Tabanus* but are usually a little longer. The third segment is straight, not curved as in *Tabanus*, has only three, not four, small subdivisions and does not bear a dorsal projection (Fig. 7.3b).

Haematopota species are not found in South America or Australasia, and only very few species occur in North America. They are common in Europe, Asia, Africa, India and the Far East.

7.2   LIFE CYCLE

Both males and females feed on naturally occurring sugary secretions, but in addition females bite a wide variety of mammals such as humans, domestic animals, especially horses and cattle, deer and many other herbivores, as well as carnivores and monkeys. A few species attack birds.

Some 100–1000 eggs, the number depending on the species, are deposited on the underside of objects such as leaves, grassy vegetation, twigs, small branches, stones and rocks. These oviposition sites overhang, or are adjacent to, the larval habitats, which are often muddy, aquatic or semi-aquatic sites (p. 112). Eggs, which are firmly glued in an upright position in a single large mass (up to 25 mm long) to the substrate, are covered with a coating that is impervious to water, i.e. they are waterproofed. They are often arranged in a more-or-less lozenge-shaped pattern (Fig. 7.5a) and are mostly creamy white, greyish or blackish. Each egg measures 1–3 mm in length and is curved or approximately cigar-shaped. Eggs hatch after 4–14 days, the time depending on both temperature and species. After wriggling out of the eggs the young larvae drop down on to the underlying mud or water.

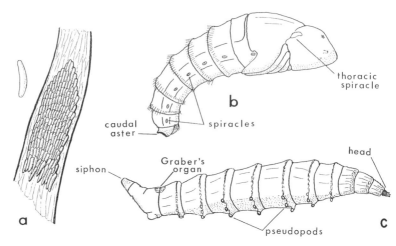

Figure 7.5 Immature stages of tabanids: (a) single egg and egg mass glued to a piece of grass; (b) pupa; (c) larva.

Larvae are cylindrical and rather pointed at both ends (Fig. 7.5c). They are creamy white, brown or even greenish but often have darkish pigmentation near the borders of the segments. There is a very small black head which can be retracted into the thorax. There are 11 well-differentiated body segments. Larvae are readily recognized by the prominent raised *tyre-like rings* which encircle most body segments. The first seven abdominal segments have one pair of lateral and two pairs of ventral (a total of six) conspicuous roundish protuberances called *pseudopods*. The presence of prominent rings and these pseudopods readily identify larvae of tabanids. The last abdominal segment has a short *siphon* dorsally which can be retracted into the abdomen, and a pyriform structure known as Graber's organ which is composed of 15 or fewer black globular bodies. This can readily be seen with the aid of a hand lens or microscope and is unique to tabanid larvae. This organ is well provided with nerves and seems to be sensory, but its exact function remains unclear.

Larvae live in mud, rotting vegetation, humus, damp soil, in shallow and often muddy waters, at the edges of small pools, swamps, ditches or slowly flowing streams. In aquatic habitats larvae sometimes adhere to floating leaves, logs or other debris. A few species breed in tree-holes. Some species occur in brackish habitats, whereas larvae of certain *Tabanus* and *Haematopota* species are found in the drier soils of pastures and in the earth near the bases of trees. Larvae breathe through the short siphon at the posterior end of the body, and because they are poor swimmers aquatic species are usually found in only shallow waters. They move rather sluggishly in their muddy, aquatic or semi-aquatic environment. Larvae of *Tabanus* and *Haematopota* are predacious or cannibalistic. Sometimes carnivorous tabanid larvae bite people, such as those working barefoot in ricefields. Such bites can be quite painful. In some species, in particular *Chrysops*, larvae are mainly scavengers, feeding on detritus and a variety of dead and decaying vegetable and animal matter.

Larval development is of a ***long duration***. In both temperate and tropical countries many species spend 1–2 years as larvae, and several cold-temperate species may remain as larvae for 3 years. Larvae are not easily found and relatively little is known about the life cycle of many species, but there appear to be 6–13 larval instars. Depending upon the species mature larvae may be 10–60 mm long. Prior to pupation mature larvae migrate to drier areas at the periphery of larval habitats where they pupate. The pupa is partially buried in the mud or soil in an upright position and, superficially, looks like a chrysalis of a butterfly (Fig. 7.5b). It is 6–35 mm long, size depending on the species, distinctly curved and is usually brown. The head and thorax are combined to form a distinct cephalothorax which has a pair of lateral and relatively large ear-shaped spiracles. The abdomen is composed of eight well-defined segments, the first seven of which are supplied with a pair of lateral spiracles, whereas segments 2–7 have an encircling row of small, backwardly directed spines. A few spines are also present ventrally on segment 8. The short terminal eighth abdominal segment is provided with six lobes which bear spine-like processes, known collectively as the ***caudal aster***. The pupal period lasts about 5–20 days.

7.2.1 Adult behaviour

Females of most species feed during the ***daytime*** and are especially active in bright sunshine, but a few species are crepuscular and some feed at night. They locate their hosts mainly by sight (colour, shape and movement), although olfactory stimuli such as carbon dioxide and other host odours also play a role in host location. Tabanids are powerful fliers and may disperse several kilometres.

Most tabanids inhabit woods and forests. Many *Chrysops* species are common in low-lying marshy scrub areas or swampy woods; some species, however, are found in more open savannah and grassland areas. Adults do not usually enter houses to feed, but *C. silaceus* is an exception. It is also attracted by smoke from wood burning and from forest fires. Other species may be found in houses, especially on windows, but often these have entered accidentally and are trying to escape.

Because of their large and rather broad mouthparts, bites from tabanids are painful, sometimes especially so, and wounds inflicted by tabanids frequently continue to bleed after the female has departed. Due to their painful bites tabanids are frequently disturbed when feeding on either people or other hosts (e.g. horses, cattle, deer, monkeys). As a result several small blood-meals may be taken from the same or different hosts before the female has obtained a complete meal. This interrupted feeding behaviour increases their likelihood of being mechanical vectors of disease. Because of their preference for dark objects (dark blue, black, red) they often bite through coloured clothing when attacking Caucasians rather than biting exposed areas of pale skin; in this respect they behave like tsetse-flies.

In both temperate and tropical areas the occurrence of adults is seasonal. In temperate countries adults usually die off at the end of the summer and a new population emerges in the spring or summer of the following year. In the tropics tabanids may not completely disappear in the dry months, but their numbers are normally much reduced. Maximum numbers of biting flies usually appear towards the beginning of the rainy season.

7.3   MEDICAL IMPORTANCE

7.3.1   Minor infections

Because females tend to be intermittent feeders, and are often disturbed during feeding, tabanids are particularly liable to be mechanical vectors. In this way they can spread anthrax (*Bacillus anthracis*), and in North America and Russia they have been incriminated in the mechanical spread of tularaemia (*Francisella tularensis*) from horses, rabbits and other rodents to humans. In North America *Chrysops discalis* has been identified as a vector of tularaemia, but other tabanid species are most likely involved. The disease is also commonly spread by handling infected rodents, by ixodid tick bites and by eating insufficiently cooked meat. It has also been suggested that Lyme disease (*Borrelia burgdorferi*), which is normally transmitted by ixodid ticks, may be transmitted by tabanids, but probably involving just mechanical transmission.

Tabanids transmit viruses, bacteria, protozoa and filarial worms to livestock and therefore are of veterinary importance.

Because of their painful bites, tabanids may sometimes constitute a pest nuisance and make outdoor activities, whether recreational or work, difficult. Some people develop severe allergic symptoms due to the large amount of saliva that is pumped into the wound to prevent blood from clotting.

7.3.2   Loiasis

The only important and cyclically transmitted disease spread to humans by tabanids is loiasis, caused by the nematode *Loa loa* which undergoes a developmental cycle in the fly. This disease occurs principally in the equatorial rain forests of Sierra Leone westwards to Ghana and then from Nigeria across Gabon to Zaire, the southern Sudan and into western parts of Uganda. The *diurnal periodic* microfilariae are more or less absent from peripheral blood of people at night but appear in it during the day, especially in the morning. The microfilariae are therefore readily picked up by *C. silaceus* and *C. dimidiatus*, species which bite during the day. In areas of

Africa such as in Bahr-el-Ghazal, Sudan, where these flies are absent, other species, such as *C. distinctipennis* and *C. longicornis*, appear to be vectors.

After taking an infective blood-meal many, but not all, ingested microfilariae survive the process of blood digestion, penetrate the gut wall and migrate to the muscular and connective tissues in the abdomen, and to a lesser extent thorax and head. Here they moult twice and develop into *third-stage* larvae (L3) (2 mm long), which migrate to the thorax and head, and after 7–15 days congregate in the proboscis. When *Chrysops* feed on humans as many as 200 L3 larvae may be deposited on the skin. Most die, but some manage to pass through the punctures made by the biting flies, or through skin abrasions, and enter the host. They migrate to the connective tissues and become mature worms in 3 months, and after a further 3 months microfilariae appear in the peripheral blood.

Loa loa has also been found in monkeys. The microfilariae appear in their peripheral blood at night and are picked up by *C. centurionis* and *C. langi*, species which are mainly crepuscular and nocturnal and which feed in the tree canopy, where monkeys are sleeping. It has sometimes been considered that there can be transmission between monkeys and humans, but this seems unlikely because the parasite found in monkeys is now regarded as a separate strain or subspecies of *L. loa*, and also because the vectors bite only in the upper forest canopies.

7.4   CONTROL

There are very few practical measures to control tabanids. In theory efficient drainage not only to remove standing water but to dry out marshy and muddy areas might reduce the numbers breeding in these habitats, but the cost of both locating the larval habitats and carrying out drainage obviates this approach. Similarly, because of the difficulty of locating breeding places and their diffuse and large size, it is usually impossible to achieve control by the application of insecticides. Moreover, because larvae of many species live below the surface of the ground, heavy dosage rates would be needed for the insecticide to penetrate through the surface soil and vegetation to reach the larvae; these problems are somewhat similar to those encountered in the control of ceratopogonid larvae (Chapter 6).

Some level of local control can sometimes be achieved by employing attractant traps such as coloured screens coated with adhesives to catch adults. But there are no really effective control methods against tabanids. Some degree of personal protection may, however, be obtained by using insect repellents.

FURTHER READING

Anderson, J. F. (1985) The control of horse flies and deer flies (Diptera: Tabanidae). *Myia*, **3**, 547–98.

Anthony, D. W. (1962) Tabanids as disease vectors. In: *Biological Transmission of Disease Agents*, ed. K. Maramorosch. Symposium held under the auspices of the Entomological Society of America, Atlantic City, 1960. New York, NY: Academic Press. pp. 93–107.

Chippaux, J.-P., Bouchité, B., Demanov, M., Morlais, I. and LeGoff, G. (2000) Density and dispersal of the loiasis vector *Chrysops dimidiata* in southern Cameroon. *Medical and Veterinary Entomology*, **14**, 339–44.

Duke, B. O. L. (1972) Behavioural aspects of the life cycle of *Loa*. In: *Behavioural Aspects of Parasite Transmission*, ed. E. K. Canning and C. A. Wright. London: Academic Press. pp. 97–107.

Foil, L. D. (1989) Tabanids as vectors of disease agents. *Parasitology Today*, **5**, 88–95.

Krinsky, W. L. (1976) Animal disease agents transmitted by horse flies and deer flies (Diptera: Tabanidae). *Journal of Medical Entomology*, **13**, 225–75.

Noireau, F., Nzoulani, A., Sinda, D. and Itoua, A. (1990) Transmission indices of *Loa loa* in the Chaillu Mountains, Congo. *American Journal of Tropical Medicine and Hygiene*, **43**, 282–8.

8

Tsetse-flies (Glossinidae)

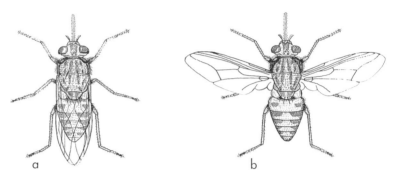

Figure 8.1 Tsetse-flies: (a) wings folded over the body like a pair of closed scissors; (b) wings purposely spread out to display abdomen and wing venation.

There are 23 species of tsetse-flies, six of which are divided into two or more subspecies, all belonging to the genus *Glossina*, the only genus in the family Glossinidae. Apart from two localities in the Arabian peninsula, tsetse-flies are restricted to sub-Saharan Africa from approximately latitude 10° north to 20° south, but extending to 30° south along the eastern coastal area. Some species, such as *Glossina morsitans*, are found across West Africa to Central and East Africa, whereas others are more restricted in their distribution. For example, *G. palpalis* occurs in only the West African subregion.

Tsetse-flies are vectors of both human and animal African trypanosomiasis, the disease in humans being referred to as sleeping sickness. The most important vectors are *G. palpalis*, *G. tachinoides*, *G. fuscipes*, *G. pallidipes* and *G. morsitans*.

8.1   EXTERNAL MORPHOLOGY

A general description of tsetse-flies, without special reference to any particular species, is as follows. Adults are yellowish or brown–black robust flies that are rather larger (6–14 mm) than house-flies. In some species the abdominal segments are uniformly coloured whereas in others they may have lighter-coloured transverse stripes and a median longitudinal one. Tsetse-flies are readily distinguished from all other biting flies and similar-sized non-biting flies by the combination of a rigid and forwardly-projecting **proboscis** and a characteristic wing venation. Between veins 4 and 5 there is a closed cell which, with a little imagination, looks like an upside-down hatchet (i.e. axe, cleaver or chopper), and is consequently often termed the **hatchet cell** (Figs. 8.1b, 8.2a). This one character serves to identify a fly as a tsetse. Tsetse-flies also differ from most flies in that at rest the **wings** are placed over the abdomen like the closed blades of a pair of scissors (Fig. 8.1a).

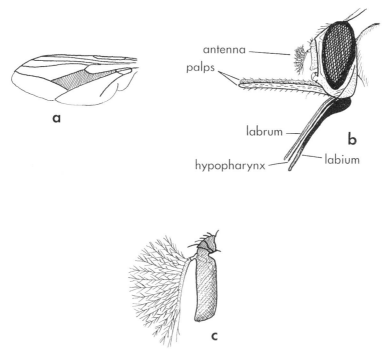

Figure 8.2 Adult tsetse-fly: (a) wing showing the hatchet cell identified by shading; (b) lateral view of head showing mouthparts, antenna and palps; (c) antenna showing plumose branching of hairs on upper side of the arista.

A long pair of palps arise dorsally, very close to the proboscis, and lie alongside it. They are difficult to distinguish except when the tsetse-fly is feeding and the proboscis is swung downwards while the palps remain projecting forwards (Fig. 8.2b). The first two antennal segments are small and inconspicuous but the third is relatively large and cylindrical and is also somewhat banana-shaped. This has near its base the arista which has branched hairs, but only on the upper surface, giving it a feathery appearance (Fig. 8.2c).

The proboscis is relatively large and has a bulbous base. When a tsetse-fly feeds saliva containing anticoagulants is pumped down into the wound formed by the fly.

The dorsal surface of the thorax of a tsetse-fly has a pattern of dark brown stripes or patches. There are eight abdominal segments and these may be uniformly dark brown or blackish, or have transverse stripes and a median one of a lighter brown or yellowish colour. Although, as far as disease transmission is concerned, it is not important to distinguish the sexes because *both* take blood-meals, this can nevertheless be done easily

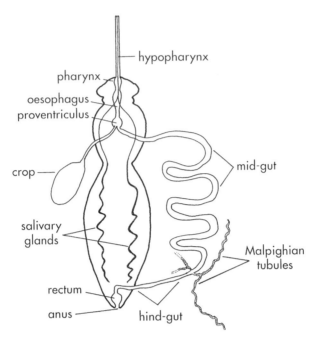

Figure 8.3 Adult tsetse-fly, dissected to show diagrammatically the alimentary canal and paired thread-like salivary glands.

by examining the tip of the ventral surface of the abdomen. In the male tsetse-fly there is a prominent raised, almost circular, button-like structure called the hypopygium, which when unfolded reveals a pair of genital claspers. In the female fly there is no such button-like protuberance.

8.1.1   Alimentary canal of the adult fly

Knowledge of the morphology of the alimentary canal and associated salivary glands (Fig. 8.3) is essential for an understanding of the life cycle of trypanosomes within the tsetse.

The food channel, formed by the apposition of the labrum and labium, leads to the pharynx and then to the oesophagus which has a slender duct leading to an oesophageal diverticulum, commonly called the crop. Just behind the oesophagus there is an important bulbous structure termed the *proventriculus*. The distal end of the proventriculus marks the end of the fore-gut and beginning of the mid-gut, which in the tsetse-fly is very long and convoluted. A *peritrophic membrane*, which plays a part in the cyclical development of sleeping sickness trypanosomes (*Trypanosoma brucei gambiense*, *T. brucei rhodesiense*) in the tsetse-fly, is secreted by the epithelial cells in the anterior part of the proventriculus. The peritrophic membrane when first produced by the proventriculus is a very delicate, soft and almost fluid structure, but as it passes back into the gut it hardens to form a thin but relatively tough sleeve, something like a sausage skin. This tube-like peritrophic membrane lines the entire length of the mid-gut.

The junction of the four Malpighian tubules separates the mid-gut from the hind-gut, which terminates in a small dilated rectum and opens to the exterior through the anus.

Paired salivary glands originating in the head of tsetse-flies are enormously long, thread-like, very convoluted and stretch back to almost the end of the abdomen. Anteriorly, the ducts from both glands unite in the head to form the common salivary duct which passes down the length of the hypopharynx, which runs down the centre of the proboscis.

8.2   LIFE CYCLE

8.2.1   Feeding and reproduction

Both male and female tsetse-flies bite people, a large variety of domesticated and wild mammals, and also reptiles and birds. No species of tsetse feeds exclusively on one type of host but most species show definite host preferences. For example, in East Africa *Glossina swynnertoni* feeds mainly on wild pigs and *G. morsitans* on wild and domesticated bovids as well as on wild pigs, but in West Africa *G. morsitans* feeds mainly on warthogs. In East Africa *G. pallidipes* feeds principally on wild bovids, while in West Africa *G. palpalis* feeds predominantly on reptiles and humans, and in West Africa *G. tachinoides* feeds on humans and bovids, although in southern Nigeria it feeds predominantly on domestic pigs. Tsetse-flies take blood-meals about every 2–3 days, although usually more frequently in dry, hot weather, while in cool humid conditions feeding intervals may be prolonged to about 10 days. Feeding is restricted to daylight hours and vision is important in host location, dark moving objects being particularly attractive. On pale-skinned people, such as Caucasians, tsetse-flies often bite through dark clothing such as socks, trousers and shorts in preference to settling on the skin. During feeding blood is sucked up the proboscis, passes to the crop and later to the mid-gut where digestion proceeds.

The different types of flies so far described in this book lay eggs; in marked contrast tsetse-flies do not, and instead they deposit larvae, one at a time (i.e. they are *larviparous*). Adults of *Sarcophaga* and *Wohlfahrtia* also deposit larvae, not eggs (Chapter 10).

After a female tsetse-fly has been inseminated by a male and after it has taken a blood-meal, a single egg in one of the two ovaries completes maturation. It then passes down the common oviduct into the uterus where it is fertilized by the release of sperm from the paired spermathecae. The egg hatches within the uterus after about 3–4 days, and the empty eggshell passes out through the genital orifice (vagina). The uterus is supplied with a conspicuous pair of branched secretory accessory glands which in tsetse-flies are called the *milk glands* (Fig. 8.4). Fatty, nutrient fluid from these

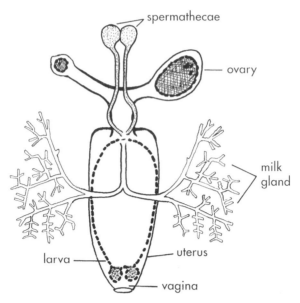

Figure 8.4 Diagrammatic representation of the female reproductive system of a tsetse-fly with a full-grown larva in the uterus.

glands flows through a small duct to enter the anterior end of the uterus. Secretions from these glands provide the larva with all the food it needs for growth and development. The larva passes through three instars in the female. Regular blood-meals must be taken by the female for a continuous and adequate provision of nutrient fluid from the milk glands. If the fly is unable to feed, the larva may fail to complete its development and as a consequence be 'aborted'.

Larval development is completed after about 9 days, by which time the third and final instar larva is 5–7 mm long. It is creamy white and composed of 12 visible segments, the last of which bears a pair of prominent dark protuberances called the ***polypneustic lobes*** (Fig. 8.5b), which are respiratory structures. A female containing a fully developed larva is easily recognized because the fly's abdomen is enlarged and stretched, i.e. the fly is obviously 'pregnant'. Furthermore, the black polypneustic lobes can be seen through her abdominal integument.

The mature third-instar larva wriggles out, posterior end first, from the genital orifice; thus birth can be termed a 'breech case'. Females always select shaded sites for larviposition. The larva is deposited on loose friable soil, sand or humus, frequently underneath bushes, trees, logs, rocks, between buttress roots of trees, in sandy river beds, in animal burrows and even in rot-holes in trees which may be formed some distance above the ground (4–5 m). Immediately the larva is deposited it commences to bury itself under 2–5 cm of soil. After about 15 minutes the third-instar larval

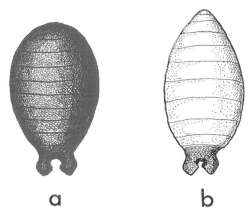

Figure 8.5 Tsetse-fly: (a) puparium; (b) larva.

skin contracts and hardens to form a reddish-brown or dark brown, barrel-shaped *puparium* which is about 5–8 mm long and has distinct polyp-neustic lobes (Fig. 8.5a). Within this puparial case the larva pupates.

The duration of the puparial period is comparatively long, usually extending over 4–5 weeks, but at high temperatures (30 °C) it may be completed within 3 weeks, and conversely at low temperatures (20 °C) prolonged to 7 weeks. After puparial development has been completed the fly emerges from the puparium, forces its way to the surface of the ground and flies away.

During development of the larva within the female the tsetse-fly feeds several times, about once every 2–3 days. The first larva is deposited about 16–20 days after the female has emerged from the puparium; thereafter, if food is plentiful a larva is deposited about every 9–12 days. In the laboratory female tsetse-flies have produced up to 20 offspring but the average is nearer 5–8. Breeding generally continues throughout the year but in very humid conditions reproduction may be diminished. Maximum population size is usually attained at the end of the rainy season. The population diminishes in the dry season when suitable areas of refuge for adult flies and suitable larviposition sites may become restricted and localized.

8.2.2   Adult behaviour

Knowledge of the behaviour of tsetse-flies is essential for understanding control strategies, and also the part species play in the transmission of sleeping sickness.

Blood-engorged tsetse-flies, and unfed hungry flies waiting to feed, spend the nights and much of the day resting in dark and usually humid sites. In fact tsetse-flies spend about *23 hours* a day resting on vegetation. During the day the favoured resting sites of most species are twigs, branches and trunks of trees and bushes. Flies are not found resting in sites in which temperatures rise above about 36 °C. At night tsetse-flies prefer to rest on the upper surfaces of leaves. Accurate knowledge of the actual resting sites may be required for control measures. For example, the height at which adults rest on trees determines the height at which the trees need to be sprayed with insecticides. Most species in fact rest below 4 m; in Nigeria 50% of *G. palpalis* and *G. morsitans* commonly rest between ground level and 30 cm.

Based on their morphology, ecology, karyotype and behaviour, tsetse-flies can be separated into the following three main groups.

Fusca group (forest flies)

The *fusca* group contains 15 species and subspecies of *Glossina*, all of which are large (10.5–15.5 mm). They are forest flies (except *G. longipennis*, which occurs in arid areas of East Africa) and most are restricted to the equatorial forests of West and western Central Africa. For example, *G. fusca* occurs mainly in relict forests of West and Central Africa, whereas *G. brevipalpis* is found in secondary forests of East Africa.

The *fusca* group rarely feeds on people and none of the species is a vector of sleeping sickness.

Morsitans group (savannah flies)

Seven species and subspecies are included within the *morsitans* group. They are medium-sized insects, 7.5–11 mm long, and typically inhabit savannah regions of Africa, which may extend from the coast, or the edges of forests, to dry semidesert regions. *Glossina morsitans* occupies the savannah regions of West, Central and East Africa, whereas *G. pallidipes* is found in savannahs of East Africa and parts of southern Africa, and *G. swynnertoni* is restricted to the savannahs of a very limited area of East Africa. Within savannahs *G. morsitans* and *G. pallidipes* occur in areas ranging from wooded sites at the edges of forests to dry thicket vegetation of arid zones, whereas *G. swynnertoni* is restricted mainly to relatively dry thicket country.

All three species above are vectors of sleeping sickness, but the most important is *G. morsitans*.

Palpalis group (riverine and forest flies)

Nine tsetse species and subspecies are included in the *palpalis* group, the smallest being about 6.5 mm in length and the largest 11 mm. They are essentially flies inhabiting wetter types of vegetation, such as forests, luxuriant scrub and vegetation growing along rivers and shores of lakes. *Glossina palpalis* inhabits riverine vegetation bordering rivers and lakes, mangrove swamps and forested areas, and occurs throughout most of West Africa, down the western part of the continent to Angola. *Glossina fuscipes*, which is closely related to *G. palpalis*, occurs mainly in Central Africa but extends its range to the western areas of East Africa. *Glossina tachinoides* is a riverine species found near streams and rivers in wet humid coastal areas, through wooded savannah regions to the riverine vegetation of very dry savannah areas. It is found mainly in West and Central Africa but also occurs in parts of Ethiopia and Sudan. All these species are vectors of sleeping sickness.

8.3   MEDICAL IMPORTANCE

All species of *Glossina* are potential vectors of African trypanosomiasis to humans. In practice, however, relatively few species of tsetse-flies are natural vectors because many species rarely, if ever, feed on people. It is the behaviour of the adult tsetse-fly and the degree of fly–human contact, and in the case of Rhodesian sleeping sickness also the degree of vector contact with reservoir hosts of the trypanosomes, that establishes whether a tsetse-fly is a vector.

Sleeping sickness has a patchy distribution over about 10^6 km^2 of land (that is an area greater than the USA) in 36 African countries. About 20 000–25 000 new cases are recorded annually but the WHO estimates there are actually about 400 000 new cases a year and 55 000 deaths. There are two subspecies of trypanosomes (previously often regarded as distinct species) causing sleeping sickness in humans, namely *Trypanosoma brucei gambiense* and *T. brucei rhodesiense*. These parasites are morphologically indistinguishable but produce different clinical symptoms and have different epidemiologies. The most important vectors of sleeping sickness are *G. palpalis*, *G. fuscipes*, *G. tachinoides*, *G. morsitans* and *G. pallidipes*.

The cycle of development of *Trypanosoma brucei gambiense* and *T. brucei rhodesiense* in the tsetse-fly is the same and is as follows. Trypanosomes in the blood are sucked up by male or female tsetse-flies during blood-feeding from an infected person (or, particularly in the case of *T. brucei rhodesiense*, often from a non-human reservoir host) and pass first through the oesophagus to the crop, and then, after feeding has ceased, into the **peritrophic tube** lining the mid-gut. Blood is digested within the mid-gut but the *gambiense* and *rhodesiense* trypanosomes are not destroyed. About 9–11 days after feeding the trypanosomes penetrate the middle section of the peritrophic membrane and pass across into the space between the membrane and gut, called the **ectoperitrophic space**. Here they multiply and after about 3–9 days the parasites penetrate the anterior softer part of the peritrophic membrane and migrate to invade the **proventriculus**. From here they pass down the food channel in the proboscis and pass up the hypopharynx or salivary duct to invade the **salivary glands**, where they develop into **epimastigotes** and multiply enormously. About 18–35 days after an infective blood-meal the tsetse-fly becomes infective and meta-cyclic **trypomastigotes** (metatrypanosomes) are injected into a vertebrate host when the fly feeds. In the 1970s there was evidence for a more direct route in which parasites passed through the peritrophic membrane and gut wall to enter the haemocoel (body cavity) and then invaded the salivary glands. It remains unclear how important this shorter route is.

When determining infection rates in tsetse-flies by dissection, any trypanosomes found in the gut or proboscis are ignored as only those in the salivary glands can be ascribed to *T. brucei gambiense* or *T. brucei rhodesiense*, but there are complications: *T. brucei brucei*, which does not cause sleeping sickness in humans but causes an animal trypanosomiasis commonly called nagana, undergoes a similar cyclical development in the fly. Consequently, the presence of trypomastigotes (metacyclic forms) in the salivary glands does not necessarily indicate the presence of trypanosomes infective to people.

Salivary gland infection rates in tsetse-flies are low, rarely exceeding 0.1%, even in endemic areas.

8.3.1   Gambian sleeping sickness

Gambian sleeping sickness (*T. brucei gambiense*) is a form of the disease that occurs from West Africa through Central Africa to parts of Sudan and southwards to Angola and Zaire. *Glossina palpalis* and *G. tachinoides* are the most important vectors in West Africa, while *G. fuscipes* is the vector in Central and East Africa. This disease is relatively chronic, with death often not occurring until after many years. Until recently it was considered that there were no natural reservoir hosts of the infection other than humans, but studies in West Africa indicate that domestic pigs and some wild mammals may harbour *T. brucei gambiense*.

Vectors of Gambian sleeping sickness are especially common at watering places, fords across rivers and along lake shores, etc., in fact in places where people frequently visit to collect water or do their washing. As a consequence there may be limited and localized foci of transmission.

8.3.2   Rhodesian sleeping sickness

The causative agent of Rhodesian sleeping sickness, *T. brucei rhodesiense*, causes a more virulent disease than *T. brucei gambiense*, with death occurring after just weeks or months. However, it is not so widespread, being more or less restricted to Tanzania, Malawi, Zambia, Zimbabwe, Mozambique and the northern areas of Lake Victoria in Kenya and Uganda. The most important vectors are *G. morsitans* and *G. pallidipes*, species which feed on a variety of game animals and domestic livestock, especially bovids, in preference to people. These flies often occur in savannah areas thinly populated by humans. Wild animals, especially various bovid species, are important reservoir hosts of *T. brucei rhodesiense*. Around Lake Victoria, *G. fuscipes* is the main vector and cattle are important reservoir hosts. This form of sleeping sickness is a ***zoonosis***.

8.4   CONTROL

Because the immature stages are so well protected, the larva being retained by the female for almost all of its life, and the puparium being buried in the soil, control of tsetse-flies is aimed at the adults.

Many control methods have been directed at tsetse-flies to combat human and animal trypanosomiasis. Formerly there were campaigns in Zimbabwe and some East African countries to kill in selected areas all game animals that might provide food for tsetse flies, or be reservoir hosts of trypanosomiasis. Wide-scale and often indiscriminate slaughter of animals is no longer acceptable in a world that is increasingly sensitive to the preservation of wildlife.

The distribution and abundance of tsetse-flies is largely determined by types of vegetation and microclimate. Their dependence on such requirements formed the basis of control through clearing away vegetation, especially near rivers and lake shores, either completely or partially. The destruction of such vegetation in the past has achieved considerable success in controlling tsetse-flies, but in many countries this method, at least on a large scale, is no longer ecologically or economically acceptable. However, clearing vegetation for human settlements or felling trees for firewood may also have the same effect in reducing tsetse-fly populations.

8.4.1   Insecticidal control

There are no problems with insecticide resistance in tsetse-flies. Consequently insecticides such as endosulfan, permethrin, deltamethrin, lambda-cyhalothrin and cypermethrin can be sprayed onto vegetation harbouring adult flies. Success depends on a detailed knowledge of the resting sites of the vectors. For example, it is often possible to restrict spraying of tree trunks up to a height of 1.5 m during the dry season, extending this to 3.5 m in the wet season. The frequency of application depends much on local conditions, such as rainfall, other climatic factors and the growth and spread of new vegetation, which can provide new resting sites free of insecticides. Spraying is often done in the dry season so that deposits are not washed from vegetation by rainfall and applications can remain effective for at least 2–3 months.

Alternatively, ultra-low-volume (ULV) aerial spraying of residual insecticides, using fixed-wing aircraft or helicopters, can be undertaken. Repeated applications, such as about five sprayings each separated by 10-day intervals, are necessary because the tsetse-fly population is in the soil as puparia for 4–5 weeks or longer, and remains unaffected by spraying. Although aerial spraying results in greater coverage of vegetation with insecticides than is achieved with ground applications, it usually kills a greater number of non-target insects and animals, many of which may be directly or indirectly beneficial to the human population.

Although indiscriminate and repeated spraying of residual insecticides can have disastrous effects on the local fauna it should be appreciated that application rates of insecticides for tsetse-fly control are often very low. Furthermore, after flies have been controlled and spraying stopped, many, if not all, of the wildlife may revert to their original population numbers. Intensive bush clearance for tsetse-fly control can also greatly diminish the numbers and variety of the local fauna, but in this instance the change may be permanent, especially if people occupy and farm areas that were originally scrub or forest.

8.4.2 Targets and traps

Although insecticidal spraying is still used to control tsestse-fly vectors of animal trypanosomiasis (nagana) there is increasing use of more environmentally friendly, and often cheaper, techniques such as employing targets and traps to kill vectors of sleeping sickness. Targets, sometimes called screens, made of dark blue or black cloth and about 1 m^2 are mounted on poles or hung from trees. They are very attractive to tsetse-flies and when they are impregnated with pyrethroids (e.g. deltamethrin, lambda-cyhalothrin or cyfluthrin) they kill adult flies settling on their surfaces. Re-impregnation is needed every 3–4 months. Newer methods still under field trials involve treating targets with insect growth regulators or chemosterilants.

Biconical or pyramidal traps made partially from dark blue or black material can be remarkably effective in attracting and trapping flies. Because some flies fail to enter traps they are often impregnated with pyrethroid insecticides to kill those resting on the outside. Traps should be retreated with insecticides about every 6–10 months. The location of both screens and traps is critical for efficient control; generally open and sunny sites are best as they offer good visibility for the flies.

Recently there has been a move away from government-organized tsetse control campaigns to operations, using screens and traps, implemented by local people. Success of such community participation for the control of sleeping sickness has been mixed; it has usually proved difficult to sustain initial enthusiasm and success after outside inputs are removed.

8.4.3 Genetic control

Although genetic control, mainly using sterile male release methods (see pp. 24–25), has sometimes been successful against tsetse-fly vectors of cattle trypanosomiasis, even leading to the eradication of flies from Zanzibar, the approach is not generally feasible for sustained control of human sleeping sickness. An exception might be to use genetic methods to eradicate residual populations of flies left locally after cheaper control methods (e.g. insecticides, traps) have drastically reduced their numbers but failed to eradicate the flies.

FURTHER READING

Buxton, P. A. (1955) *The Natural History of Tsetse-Flies: an Account of the Biology of the Genus* Glossina (*Diptera*). London School of Hygiene and Tropical Medicine Memoir, 10. London: H. K. Lewis.

Challier, A. (1982) The ecology of tsetse (*Glossina* spp.) (Diptera: Glossinidae): a review (1970–1981). *Insect Science and its Application*, **3**, 97–143.

Colvin, J. and Gibson, G. (1992) Host-seeking behavior and management of tsetse. *Annual Review of Entomology*, **37**, 21–40.

Dransfield, R. D., Williams, B. G. and Brightwell, R. (1991) Control of tsetse flies and trypanosomiasis: myth or reality? *Parasitology Today*, **7**, 287–91.

Ford, J. (1971) *The Role of Trypanosomiases in African Ecology: a Study of the Tsetse Fly Problem*. Oxford: Clarendon Press.

Green, C. H. (1994) Bait methods for tsetse fly control. *Advances in Parasitology*, **34**, 229–91.

Jordan, A. M. (1986) *Trypanosomiasis Control and African Rural Development*. London: Longman.

(1989) Man and changing patterns of the African trypanosomiases. In: *Demography and Vector-Borne Diseases*, ed. M. W. Service. Boca Raton, FL: CRC Press. pp. 47–58.

(1993) Tsetse-flies (Glossinidae). In: *Medical Insects and Arachnids*, ed. R. P. Lane and R. W. Crosskey. London: Chapman and Hall. pp. 333–88.

Laveissière, C., Vale, G. A. and Gouteaux, J. P. (1991) Bait methods for tsetse control. In: *Appropriate Technology in Vector Control*, ed. C. F. Curtis. Boca Raton, FL: CRC Press. pp. 47–74.

Leak, S. G. A. (1999) *Tsetse Biology and Ecology: Their Role in the Epidemiology and Control of Trypanosomiasis*. Wallingford: CABI.

Mulligan, H. W. (ed.) (1970) *The African Trypanosomiases*. London: Allen and Unwin.

Nash, T. A. M. (1969) *Africa's Bane: the Tsetse Fly*. London: Collins.

World Health Organization (1997) *Vector Control: Methods for Use by Individuals and Communities*, prepared by J. A. Rozendaal. Geneva: World Health Organization.

(1998) Control and surveillance of African trypanosomiasis. *World Health Organization Technical Report Series*, **881**, 1–113.

9

House-flies and stable-flies (Muscidae) and latrine-flies (Fanniidae)

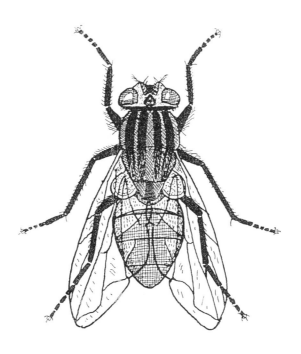

9.1   MUSCIDAE: HOUSE-FLIES AND STABLE-FLIES

There are about 4200 species of flies in 170 genera belonging to the family Muscidae. The medically most important are the common house-fly, other closely related flies and the stable-fly, all having a more-or-less world-wide distribution. House-flies can be vectors of helminths, faecal bacteria, protozoans and viruses, resulting in the spread of such enteric diseases as the dysenteries and typhoids, whereas the stable-fly can cause a biting nuisance.

9.2   THE COMMON HOUSE-FLY (*MUSCA DOMESTICA*)

9.2.1   External morphology

There are about 70 species of flies in the genus *Musca*. The most common is *M. domestica*, the house-fly, which occurs in all parts of the world but is least common in Africa, where it is largely replaced by two subspecies (*M. domestica curviforceps* and *M. domestica calleva*). Other important related species are (1) the bazaar-fly (*Musca sorbens*), which can be a great nuisance in Africa, Asia and the Pacific; (2) the notorious troublesome bush-fly of Australia, namely *M. vetustissima*; and (3) the face-fly (*M. autumnalis*), which is a pest in both the Old and New Worlds. The appearance and biology of these *Musca* species are very similar. The morphology and biology of the house-fly (*M. domestica*) are described here.

House-flies are medium sized ***non-metallic*** flies about 6–9 mm in length, varying in colour from light to dark grey. They have four broadish black longitudinal stripes on the dorsal grey surface of the thorax (Fig. 9.1a). The antennae are concealed in depressions on the front of the face of the fly and are not easily seen, but consist of three segments, the distal and largest of which is cylindrical and has a prominent hair, called an ***arista***, which has hairs on ***both*** sides. The mouthparts (proboscis) of the house-fly are specially adapted for sucking up fluid or semifluid foods. When not in use they are partially withdrawn into the head capsule (Fig. 9.2a), but are extended vertically downwards in a telescopic fashion when the fly feeds (Fig. 9.2b). The proboscis ends in a pair of oval-shaped fleshy *labella*, having very fine channels called *pseudotracheae* through which fluids are sucked up. House-flies feed on a great variety of substances, including almost all food of humans, rotting vegetables, carcasses, excreta and vomit – in fact almost any organic material. The method of feeding differs according to the physical state of the food. For example, for thin fluids, such as milk and beer, the labella are closely adpressed on the food which is then sucked up through the small openings in the pseudotracheae. When feeding on semisolids such as excreta, sputum and nasal discharges, the labella are completely everted and food is sucked up directly into a food channel formed by the apposition of the slender labrum and blade-like hypopharynx. If flies feed on more solid materials such as sugar lumps,

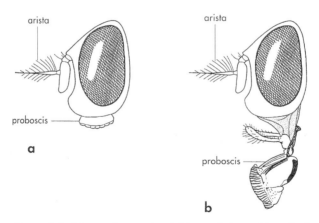

Figure 9.1 House-fly (*Musca domestica*): (a) adult fly; (b) terminal tarsal segments showing paired claws, paired large pulvilli and a single bristle-like empodium.

Figure 9.2 *Musca domestica*: (a) lateral view of head of adult house-fly with proboscis retracted; (b) proboscis extended for feeding.

dried blood, cheese and cooked meats, the labella are everted and minute prestomal teeth surrounding the food channel are exposed and scrape away at the solid food. The fly then moistens small food particles with either saliva or the regurgitated contents of its crop, after which the food is sucked up. This latter type of feeding is clearly conducive to the spread of a variety of pathogens.

The wings of the house-fly have *vein 4* bending up sharply to join the costa (the thick vein along the front edge of the wing) close to vein 3

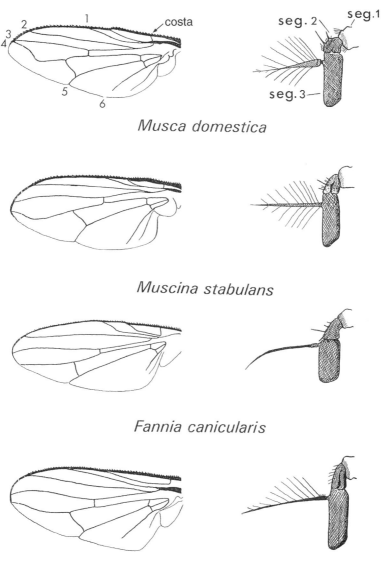

Figure 9.3 Wing venation and antennae characteristic of the genera *Musca*, *Muscina*, *Fannia* and *Stomoxys*. Note endings of veins 3 and 4.

(Fig. 9.3). This is an important taxonomic character which can help distinguish *Musca* species from other rather similar flies. All three pairs of legs end in paired claws and a pair of fleshy pad-like structures called the ***pulvilli***, which are supplied with glandular hairs (Fig. 9.1b). These sticky hairs enable the fly to adhere to very smooth surfaces, such as windows, and are also responsible for the fly picking up dirt and pathogens when it visits excreta, septic wounds, rubbish dumps, etc. There are four

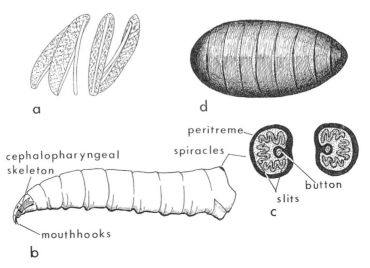

Figure 9.4 *Musca domestica* (house-fly): (a) two hatched and two unhatched eggs; (b) final-instar larva; (c) posterior larval spiracles showing their D-shape; (d) puparium.

visible greyish abdominal segments which are usually partially obscured from view by the wings.

9.2.2   Life cycle

Female *Musca domestica* lay their eggs on a variety of decomposing materials such as horse manure, poultry dung, urine-contaminated bedding, carcasses, decomposing organic materials found in rubbish dumps, household garbage and waste foods from kitchens and hotels. Some 75–120 eggs are deposited together, or in separate batches. A fly may lay eggs five or six times in her lifetime. Eggs are creamy-white, 1–1.2 mm long, and distinctly concave dorsally, giving them a banana-shaped appearance (Fig. 9.4a). They can hatch after only 6–12 hours, but this period is extended in cool weather. Hatching is accomplished by the strip of eggshell between parallel ridges on the dorsal concave surface lifting up, and partially detaching itself from the rest of the egg. Eggs cannot withstand desiccation and die if they dry out. Neither can they tolerate extremes of temperatures, most dying after exposure to temperatures below 15 °C or above 40 °C.

Larvae, known as *maggots*, which hatch from the eggs have a small head followed by an 11-segmented cylindrical body (Fig. 9.4b). At the pointed head end beneath the integument there is a pair of blackish small curved mouthhooks, which are continued as the cephalopharyngeal skeleton in the anterior thoracic segments. At the posterior end of the body there is a pair of conspicuous *spiracles* shaped like a *letter D*. Each spiracle has a complete and thick outer wall called the *peritreme* which encloses three very sinuous spiracular slits (Fig. 9.4c).

Larvae feed on liquid food resulting from decomposing organic material. There are three larval instars. Mature larvae measure about 8–14 mm, the final size depending on environmental conditions, especially the amount of available food. The speed of larval growth and development depends on the abundance of food supply and temperature. Development may be completed within just 3–5 days, but under less favourable conditions 7–10 days are needed, and in cool weather development may extend to about 24 days. If larval habitats dry out the larvae are killed, and if they become too wet they drown.

Prior to pupation third-instar larvae may move away from their larval habitats to drier ground. Sometimes, however, the periphery of breeding places, such as rubbish dumps, may be sufficiently dry for larvae to pupate there; pupation may also occur in the dry soil underneath larval habitats. Pupation begins with the larval skin contracting, hardening and turning dark brown, resulting in the formation of a barrel-shaped structure measuring about 6 mm, called a *puparium*. Close examination shows this is segmented (Fig. 9.4d). The puparium is often referred to as the pupa but technically this is incorrect because the actual pupa is formed within the protective shell of the puparial case. A puparium is also formed by tsetse-flies (Chapter 8) and by myiasis-causing flies (Chapter 10). The puparial stage lasts about 3–5 days in warm weather but 7–14 days during cooler periods.

Developmental time from egg to adult (Fig. 9.5) is about 49 days at 16 °C, 25 days at 20 °C, 16 days at 25 °C, 10–12 days at 30 °C and 8–10 days at 35 °C. Very occasionally the period can be less than 7 days. Immature development ceases at temperatures below 12 °C and a temperature of about 45 °C is lethal to the eggs, larvae and puparia. In temperate areas a varying, but usually small, proportion of house-flies survive throughout the winter as puparia, but more frequently they overwinter as hibernating adults.

The adult fly escapes from its puparial case by pushing off its anterior end and crawling out, and after a short period flies away.

Adults of *Musca domestica* generally avoid direct sunlight, preferring to seek shelter in buildings inhabited by people or their animals. House-flies and related flies, and the calliphorid flies discussed in Chapter 10, are often called domestic or *synanthropic* flies, because of their close association with humans and their homes. House-flies defecate at random, and frequently regurgitate their food, resulting in unsightly 'fly spots'. Adults tend to stay within 500 m of their breeding sites, but may fly 1–5 km or sometimes even greater distances.

Musca sorbens is widely distributed in Africa and parts of Asia, where it is a greater pest than *M. domestica* because adults more commonly settle

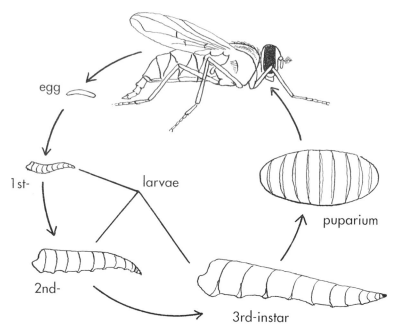

egg

1st-   larvae

puparium

2nd-

3rd-instar

Figure 9.5 Life cycle of the house-fly (*Musca domestica*), which is also typical of other muscid and calliphorid flies (see Chapter 10).

around the eyes, on sweaty skin, suppurating wounds, and other body secretions than does the common house-fly.

9.2.3   Medical importance

House-flies can transmit a large number of diseases to humans owing to their habits of visiting, almost indiscriminately, faeces and other unhygienic matter and people's food. Pathogens can be transmitted by three possible routes:

(1)   By flies' contaminated feet, body hairs and mouthparts. Most pathogens, though, remain viable on the fly for less than 24 hours, and there are usually insufficient numbers to cause a direct infection, except possibly with *Shigella*. However, if pathogens are first transferred to food they may then multiply sufficiently to reach the level of an infective dose.

(2)   By flies vomiting on food during feeding, which they do frequently.

(3)   By defecation, which often occurs on food. This is probably the most important method of transmission.

Over 100 different pathogens have been recorded from house-flies and it has been reported that 65 of them are transmitted. With the exception of *Thelazia* species (eye-worms) transmission is mechanical, that is the fly acts just as a physical carrier.

House-flies can transmit **viruses** of polio, Coxsackie and infectious hepatitis; **rickettsiae** of Q fever (*Coxiella burnetii*); **bacteria** such as anthrax, *Campylobacter*, cholera (*Vibrio cholerae*), *Shigella* and *Salmonella* species, *Escherichia coli*, *Staphylococcus aureus*, trachoma (*Chlamydia trachomatis*); **spirochaetes** of yaws (*Treponema pertenue*); **protozoans** including *Entamoeba, Cryptosporidium* and *Giardia*. In addition, house-flies can carry eggs of a variety of helminths, for example *Taenia, Hymenolepis, Dipylidium, Diphyllobothrium, Enterobius, Trichuris* and *Ascaris*. Eye-worms (*Thelazia* species) are rather rare infections of the eye and are transmitted biologically by *Musca* species. Flies also can be carriers in the tropical Americas of the eggs of *Dermatobia hominis*, a myiasis-producing fly (Chapter 10).

Larvae of house-flies have occasionally been recorded in cases of urogenital and traumatic myiasis, and more rarely in aural and nasopharyngeal myiasis. If food infected with fly maggots is eaten, then they may be passed more or less intact in the excreta, often causing considerable alarm and surprise. There is, however, no true intestinal myiasis in humans (see Chapter 10 for accounts of myiasis).

Although house-flies are potential vectors of many pathogens to humans it is difficult to assess their importance in disease transmission. Much information of their real role in the spread of disease has been circumstantial; for example, seasonal increase of fly abundance is often closely correlated with outbreaks of diarrhoeal diseases. The classic demonstration of the association between flies and disease was in two groups of Texan towns in 1946 and 1947. One group was sprayed with DDT to destroy the house-flies and this was accompanied by a reduction in acute diarrhoeal diseases and deaths in children due to *Shigella*, although *Salmonella* infections remained the same The unsprayed town did not have such a reduction in *Shigella* infections. Later the unsprayed towns were sprayed and the towns previously sprayed left unsprayed. The incidence of dysentery in the two groups was now reversed. In 1988 control of house-flies by attractant traps in an Israeli army camp resulted in a reduction in *Shigella* infections and also apparently ETEC infections. More recently when in 1995 and 1996 breeding sites in Pakistan villages were sprayed with insecticides fly populations decreased by about 97% and the incidence of childhood diarrhoea decreased by 23%. Similarly in The Gambia ultra-low-volume (ULV) spraying with deltamethrin in 1997 and 1998 reduced fly populations by about 75% and there were 75% fewer new cases of trachoma; diarrhoea in children was reduced by 25%. There is now considerable evidence that house-flies contribute to ill health.

9.2.4   Control
Control methods can be divided conveniently into three categories:

(1)   Physical and mechanical control
(2)   Environmental sanitation
(3)   Insecticidal control

Physical and mechanical control
Flies and other obnoxious insects can sometimes be prevented from entering buildings by screening windows, openings, air vents, etc. Plastic material is preferable because metal screening is liable to corrode. Mesh size of 3–4 strands per centimetre (i.e. 2–3 openings per cm) will exclude relatively large insects such as house-flies from buildings without unduly decreasing air circulation or light. Screening can be costly, but may be worthwhile in hospitals and restaurants. Screening should be periodically inspected and any tears mended. Although screening can reduce fly nuisance it does not solve the problem, because flies will continue to breed locally and to enter unscreened houses.

The establishment in doorways of an air current, such as the air barriers found in the entrances of some shops, and fans mounted over doorways, may help to reduce the number of flies entering premises. The well-known practice of placing in doorways curtains made of many vertical, often coloured, strips of plastic or beading also helps to keep out flies. A common method to kill flies that enter restaurants and food stores, is to mount an ultraviolet light-trap on a wall. Flies attracted to the trap are electrocuted on entering it by contact with an electric grid.

Commercially available sticky tapes ('fly-papers') incorporating sugar, as an attractant, can be relatively effective, although unsightly, in catching flies.

Environmental sanitation
Environmental sanitation aims at curtailing house-fly populations by reducing their breeding places. For example, domestic refuse and garbage should be placed either in strong plastic bags with the openings tightly closed, or in dustbins with tight-fitting lids. When possible there should also be regular refuse collections, preferably twice a week in warm countries, to prevent any eggs laid among the garbage developing into adults. If household refuse cannot be collected it should be burnt or buried. Unhygienic rubbish dumps, so commonly found in towns and villages, provide ideal breeding places for house-flies, and should be removed. Refuse should be placed in pits which are covered, daily if possible, with a 15-cm layer of earth; when the pits are more or less full they must be finally covered with 60 cm of compacted earth. This depth of final fill is required to prevent rodents burrowing into the rubbish pits.

It is also important to organize efficient disposal of household and industrial sewage and sanitary wastes. There should also be efficient latrines which prevent the breeding of house-flies and allied flies.

Insecticidal control
In many parts of the world house-flies have become resistant to the organochlorine insecticides, which are now rarely used. Resistance has also developed to many of the organophosphates, and consequently carbamates and the pyrethroids are increasingly used, although some house-fly populations have also become resistant to these chemicals.

> *Larvicides*. Insecticides can be directed against the larvae, such as by spraying the insides of dustbins and refuse and garbage heaps, manure piles and other breeding sites with organophosphate insecticides such as diazinon, bromophos, dimethoate or fenitrothion, or carbamates such as propoxur or carbaryl, or pyrethroids such as cypermethrin, deltamethrin or permethrin. Insect growth regulators (IGRs) such as diflubenzuron, cryomazine or pyriproxyfen are also used. Usually large volumes (0.5–5 litres/m^2) are needed to penetrate the upper 10–15 cm of breeding sites to reach the larvae.
>
> *Spraying against adults*. For indoor use commercially available aerosol spray cans or hand sprayers can be used to spray knock-down insecticides to give an immediate kill of adult flies. Suitable insecticides are organophosphates such as 2–4% malathion, 1% naled, 2% pirimiphos methyl, 0.1–0.5% fenchlorphos, or 0.1% pyrethroids (e.g. permethrin, cypermethrin, deltamethrin). Care must be taken not to contaminate food with insecticides. Aerosol applications and space-spraying have virtually no residual effects; consequently treatments have to be repeated to achieve control, and this can be costly. Furthermore this approach does little to alleviate the source of the fly nuisance.
>
> Outdoor aerosol applications from special spraying machines, or aerial ULV spraying, can give effective control in certain situations, such as in and around dairies, farms, markets and recreational areas. Organophosphate, carbamate and pyrethroid insecticides such as those used for indoor space-spraying are suitable. Usually, however, only temporary relief from flies is achieved because outdoor spraying does not usually kill indoor-resting flies, prevent reinvasion of flies from outside the treated area, or prevent the emergence of new flies from breeding places.
>
> *Residual spraying*. Flies may also be controlled by spraying indoor walls, ceilings, doors, etc. with organophosphates such as 1–2 g/m^2 of malathion, fenchlorphos, fenitrothion or pirimiphos methyl; or 0.4–1.0 g/m^2 diazinon or naled; or carbamates such as 0.1–0.2 g/m^2

bendiocarb or propoxur; or 0.025–0.1 g/m^2 cypermethrin or permethrin; or 0.01–0.15 g/m^2 deltamethrin. These residual insecticides should remain effective for 1–2 months, but much depends on local circumstances and whether the walls are washed. The outside walls of houses and cattle sheds may also be sprayed with residual insecticides, but their duration of effectiveness will depend on several factors, such as whether the surface deposit is rubbed off, or washed off by rain.

The outside of dustbins and any adjacent walls should also be sprayed to deter gravid (egg-laden) flies from ovipositing in dustbins and other nearby breeding places.

Insecticidal cords. These are cord or rope strips soaked in insecticides such as diazinon, fenthion, propoxur, cypermethrin or permethrin and dyed, preferably red, to alert people that they are impregnated with insecticides. They are hung up in dairies and other premises and flies are killed when they rest on them.

Dichlorvos strips. Dichlorvos-impregnated resin strips which disseminate insecticidal vapours are commonly used to control flies in houses, restaurants, hotel kitchens, dairies, etc. They can remain effective for 2–3 months, but this depends on temperature and degree of ventilation in the rooms in which they are used. However, because of concerns over prolonged breathing of dichlorvos vapours the sale of this insecticide has been banned in some countries, such as the UK and some states in the USA.

Toxic baits. Simple traps can be baited with an attractant, such as a bowl of fermenting yeast, and flies entering the trap prevented from escaping; in Israel such traps have caught several kilograms of flies over a few days! More usually baits consist of sugar mixed with sand, bran or some other inert carrier treated with 0.1–2% insecticides, which provides an attractive but lethal solid bait. Such baits can be scattered on the floor or placed in trays; they need to be replaced about every 2 days. Liquid baits commonly comprise 10% sugar dissolved in water plus 0.1–2% insecticides which are placed in a glass bottle inverted over a saucer-like receptacle so that as the bait evaporates more flows in from the reservoir, as in automatic feeders for poultry. The most commonly used insecticides in both dry and liquid baits are malathion, diazinon, dichlorvos, naled, fenchlorphos, trichlorphon and bendiocarb. In some countries commercially prepared dry or liquid baits are available. Occasionally paint-on baits comprising molasses, syrup or sugar, a thickening agent, and the organophosphate insecticide trichlorphon are applied directly to walls or to strips of material suspended from ceilings or fastened to walls. The house-fly sex pheromone called muscalure is sometimes added to baits to make them more attractive. Attractant baits can sometimes quickly reduce fly populations.

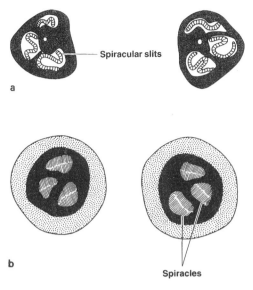

Figure 9.6 Pairs of posterior larval spiracles: (a) *Stomoxys calcitrans* (stable-fly); (b) *Muscina stabulans* (greater house-fly).

9.3   THE GREATER HOUSE-FLY (*MUSCINA STABULANS*)

9.3.1   External morphology

Muscina stabulans has a worldwide distribution and is commonly referred to as the greater house-fly, or false stable-fly. Adults are about 7–10 mm long, being slightly larger than house-flies. They are distinguished from both *Musca* and *Fannia* species because **vein 4** of the wing curves slightly, but distinctly, upwards towards vein 3 (Fig. 9.3), and from *Fannia*, but not *Musca*, by having hairs on **both** the upper and lower sides of the arista (Fig. 9.3).

9.3.2   Life cycle

Females scatter their eggs mainly on decaying organic matter and human and animal excreta. Eggs hatch after 1–2 days and the resultant larvae resemble the maggot-shaped larvae of the house-fly, but can be distinguished by the structure of the posterior spiracles. In *M. stabulans* the **spiracles** are almost **circular**, not D-shaped as in *M. domestica*. Furthermore, the **peritreme**, which is very wide, encircles three **crescent-shaped** spiracular slits (Fig. 9.6b). Young larvae are omnivorous scavengers, as are larvae of *Musca* and *Fannia*, but towards the end of the larval period they become predacious, feeding on other fly larvae. The brown puparium is similar in shape to that of *M. domestica* and its duration is about 1–2 weeks. The life cycle lasts about 4–6 weeks, although in warm weather it may be reduced to 20–25 days.

Figure 9.7 Lateral view of the head of *Stomoxys calcitrans* (stable-fly) showing the forwardly-projecting proboscis.

Adults feed and behave much the same as do adults of *M. domestica*, and may enter buildings.

9.3.3   Medical importance

Adults frequently defecate on fruit intended for human consumption, but the exact role of *M. stabulans* as a disease vector remains undetermined. However, many of the pathogens transmitted by the house-fly are probably also spread by *Muscina stabulans*. Larvae, like those of *Fannia* and *Musca*, are occasionally found in stools, but there is no true intestinal myiasis.

9.3.4   Control

Control measures are very similar to those applied to the house-fly.

9.4   THE STABLE-FLY (*STOMOXYS CALCITRANS*)

9.4.1   External morphology

Stable-flies have a worldwide distribution. They are sometimes known as biting house-flies or, in the USA, as dog-flies (because they commonly bite dogs). The most common species is *Stomoxys calcitrans*. Adults have four black longitudinal stripes on a dark grey thorax and are about the size (5–6 mm) of house-flies, which they superficially resemble. They are, however, easily separated from *Musca*, *Fannia* and *Muscina* by their conspicuous forwardly-projecting, **rigid proboscis** (Fig. 9.7). The wing venation resembles that of *Muscina*. The **arista** of the third antennal segment differs from *Musca* and *Muscina* in having hairs arising from **only** the upper side (Fig. 9.3).

In Africa adults might at first glance be confused with tsetse-flies, which also have a forwardly-projecting proboscis, but *Stomoxys calcitrans* is a smaller fly. Also, when at rest its wings are not placed completely over the body in a closed scissor-like fashion as in tsetse-flies, but are kept apart as in house-flies. Furthermore, there is no enclosed hatchet cell in the wings, as found in tsetse-flies.

9.4.2   Life cycle

Both males and females take blood-meals from wild and domesticated animals, including cattle, horses, pigs and dogs; they also feed on humans, especially if their preferred hosts are absent or scarce. During feeding the forwardly projecting proboscis is swung downwards and the skin penetrated; bites can be painful and most are on the legs. In hot weather blood-meals are digested within 12–24 hours and adults feed about every 1–3 days, but in cooler conditions blood digestion takes 2–4 days or longer, and feeding occurs every 5–10 days. *Biting* is restricted to the daytime and occurs both in bright sunshine and cloudy overcast weather. Although biting occurs mainly outdoors stable-flies will also enter houses to feed. Adults are mostly encountered in and around farms or where horses are kept, and consequently they are more common in rural areas than in towns.

The creamy-white eggs resemble those of house-flies and are usually laid in batches of less than 20, but sometimes as many as 50–200 may be laid together. Eggs are usually deposited in horse manure but also in compost pits, decaying and fermenting piles of vegetable matter, weeds, cut grass or hay. Stable-flies very rarely lay their eggs in human or animal faeces, unless these are liberally mixed with hay or straw.

Eggs hatch within 1–5 days and the resultant larvae are creamy-coloured *maggots* which resemble those of the house-fly. However, they can be separated by the arrangement of the posterior *spiracles*, which are widely separated (Fig. 9.6a), thus differing from the more closely positioned spiracular plates of *Musca* and *Muscina*. Spiracles are also approximately round in outline, lack a conspicuous *peritreme*, and the S-shaped spiracular slits are widely separated from each other. Larvae prefer a high degree of moisture for development and are found mostly in wet mixtures of manure and soil or straw, and in vegetable matter in advanced stages of decay. Under optimum conditions the larval period lasts about 6–10 days, but in cooler weather, or when there is a shortage of food, larval development can be prolonged to 4–5 weeks or more.

Larvae migrate to drier areas and bury themselves in the soil prior to pupation. The puparium is dark brown and resembles that of the house-fly, but can be distinguished from it by having the posterior spiracles widely separated. The puparial stage usually lasts 5–7 days, although in cool weather it may be extended to several weeks. The life cycle from egg-laying to adult emergence takes 12–42 days, the time depending mainly on temperature.

In tropical areas stable-flies breed continuously throughout the year, but in more temperate climates they pass the cooler months as larvae or puparia. Sometimes adults survive the winter in warm stables or buildings, feeding intermittently during the cooler months.

9.4.3   Medical importance

Because of the painful bites inflicted by *both* sexes stable-flies can be troublesome pests of people, cattle, horses and pets – notably dogs. Adult flies are not regarded as transmitting diseases to humans, and although they have been considered as mechanical vectors of African trypanosomiasis there is little evidence that they play any part in the epidemiology of sleeping sickness. Because stable-flies rarely visit excreta and festering wounds they are much less likely to spread pathogens in the way that house-flies do.

In the tropical Americas eggs of *Dermatobia hominis*, a myiasis-producing fly (Chapter 10), are sometimes attached to adult stable-flies.

9.4.4   Control

Many of the control methods aimed at house-flies can be applied with some modification to control stable-flies – for example, not allowing piles of manure, grass cuttings or decaying vegetable matter to accumulate, and burning straw and bedding material. Breeding places can be sprayed with insecticides, but as noted in the section on house-fly control it is difficult to get insecticides to penetrate deeply to where most larvae are found. Insecticidal spraying of horse stables, animal shelters, barns and other farm buildings can help reduce their numbers.

9.5   FANNIIDAE: LESSER HOUSE-FLIES AND LATRINE-FLIES (*FANNIA* SPECIES)

9.5.1   External morphology

Here the genus *Fannia* is placed in the family Fanniidae, but some consider this should be a subfamily, Fanniinae, of the family Muscidae. Flies of the genus *Fannia* resemble house-flies but are rather smaller (6–7 mm), and are readily distinguished from house-flies and other similar flies by their wing venation and antennae (Fig. 9.3). In adult *Fannia* **vein 4** of the wing is more or less parallel to vein 3, whereas in *Musca* it bends upwards and almost touches vein 3 at the wing apex, and in *Muscina* vein 4 also bends upwards but not to such an extent. The **arista**, which arises from the third antennal segment, is **completely** devoid of hairs (Fig. 9.3), whereas in both *Musca* and *Muscina* it bears branches on both sides. In most other respects adult flies are similar to house-flies. There are two common species of *Fannia* which are of minor medical importance, namely *Fannia canicularis* (the lesser, or little, house-fly), which occurs worldwide and is commonly encountered in houses, and *Fannia scalaris* (the latrine-fly), which has a Holarctic distribution and is less common in houses. Adults of *Fannia canicularis* have

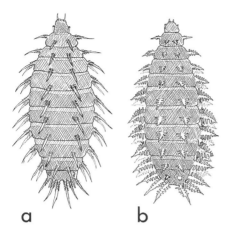

a   b

Figure 9.8 Final-instar larvae: (a) *Fannia canicularis* (lesser house-fly);
(b) *Fannia scalaris* (latrine-fly).

three rather faint dark longitudinal stripes on the thorax, whereas *F. scalaris* has *two* such stripes.

9.5.2 Life cycle

Eggs of the lesser house-fly (*F. canicularis*) resemble those of the common house-fly. They are deposited on people's food, but also on urine-soaked bedding of humans and animals, compost heaps, decaying piles of grass, human and animal excreta and in poultry litter. The latrine-fly (*F. scalaris*) usually lays her eggs in faeces – hence its common name. Larvae of *Fannia* species prefer wetter breeding places than house-flies and are often found in semiliquid habitats, such as faeces.

Eggs hatch after 1–2 days. *Fannia* larvae are quite distinct from the maggot-shaped larvae of *Musca* and are unlikely to be confused with the larvae of any other medically important fly. Figure 9.8 shows minor differences between larvae of *F. canicularis* and *F. scalaris*. They are flattened dorsoventrally and have many thin but conspicuous fleshy processes arising from the body segments which bear small spiniform secondary processes. Larval development takes about 7–12 days but may be prolonged if habitats start to dry out. The puparium is brown and is similar in shape to the larva. After 7–10 days the adult fly emerges. The life cycle often lasts about 1 month, which is considerably longer than in *Musca domestica*, but may under favourable conditions be completed within 13–22 days. The feeding mechanism of the adult is similar to that of *M. domestica*. Although *Fannia* adults often enter houses they do not settle on people, or on their food, as much as house-flies, but tend to hover around the head in an annoying manner.

9.5.3   Medical importance

Many of the pathogens transmitted by house-flies are probably also spread by *Fannia* species. They have been incriminated in cases of aural and urogenital myiasis, and larvae are sometimes found in stools, but as previously stressed true intestinal myiasis does not occur in humans.

9.5.4   Control

The same control methods apply to species of *Fannia* as to *Musca*, but particular attention should be given to the prevention and eradication of *Fannia scalaris* breeding in latrines.

FURTHER READING

Bidawid, S. P., Edeson, J. F. B., Ibrahim, J. and Matossian, R. R. (1978) The role of non-biting flies in the transmission of enteric pathogens (*Salmonella* species and *Shigella* species) in Beirut, Lebanon. *Annals of Tropical Medicine and Parasitology*, **72**, 117–21.

Chavasse, D. C., Shier, R. P., Murphy, O. A., Huttly, S. R. A., Cousens, S. N. and Akhtar, T. (1999) Impact of fly control on childhood diarrhoea in Pakistan: community-randomised trial. *Lancet*, **353**, 22–5.

Cohen, D., Green, M., Block, C., Slepon, R., Ambar, R., Wasserman, S. S. and Levine, M. M. (1991) Reduction of transmission of shigellosis by control of houseflies (*Musca domestica*). *Lancet*, **337**, 993–7.

Curtis, C. (1998) The medical importance of domestic flies and their control. *Africa Health*, **20** (6), 14–15.

Echeverra, P., Harrison, B. A., Tirapat, C. and McFarland, A. (1983) Flies as a source of enteric pathogens in a rural village in Thailand. *Applied and Environmental Microbiology*, **46**, 32–6.

Emerson, P. M., Bailey, R. L., Olaimatu, O. S., Walraven, G. E. L. and Lindsay, S. W. (2000) Transmission ecology of the fly *Musca sorbens*, a putative vector of trachoma. *Transactions of the Royal Society of Tropical Medicine and Hygiene*, **94**, 28–32.

Emerson, P. M., Lindsay, S. W., Walraven, G. E. L., Faal, H., Bøgh, C., Lowe, K. and Bailey, R. L. (1999) Effect of fly control on trachoma and diarrhoea. *Lancet*, **353**, 1401–3.

Greenberg, B. (1971) *Flies and Disease. Volume 1: Ecology, Classification and Biotic Associations*. Princeton, NJ: Princeton University Press.
    (1973) *Flies and Disease. Volume 2: Biology and Disease Transmission*. Princeton, NJ: Princeton University Press.

Krafsur, E. S. and Moon, R. D. (1997) Bionomics of the face-fly, *Musca autumnalis. Annual Review of Entomology*, **42**, 503–23.

Levine, O. S. and Levine, M. M. (1991) Houseflies (*Musca domestica*) as mechanical vectors of shigellosis. *Review of Infectious Diseases*, **13**, 688–96.

Lindsay, D. R., Stewart, W. H. and Watt, J. (1953) Effect of fly control on diarrheal diseases in an area of moderate morbidity. *Public Health Reports, Washington DC*, **68**, 361–7.

Olsen, R. A. (1998) Regulatory action criteria for filth and other extraneous materials: III. Review of flies and foodborne enteric diseases. *Regulatory Toxicology and Pharmacology*, **28**, 199–211.

Skidmore, P. (1985) *The Biology of the Muscidae of the World*. Series Entomologica 29. Dordrecht: W. Junk.

Sulaiman, S., Sohadi, A. R., Yurms, H. and Iberahim, R. (1988) The role of some cyclorrhaphan flies as carriers of human helminths in Malaysia. *Medical and Veterinary Entomology*, **5**, 72–80.

West, L. S. (1951) *The Housefly: its Natural History, Medical Importance and Control*. Ithaca, NY: Comstock Publishing; Cornell University Press.

Zumpt, F. (1973) *The Stomoxyine Biting Flies of the World. Diptera: Muscidae. Taxonomy, Biology, Economic Importance and Control Measures*. Stuttgart: Gustav Fischer.

10

Flies and myiasis

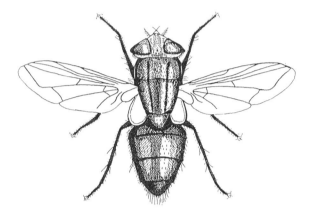

Myiasis can be defined as the invasion of organs and tissues of humans or other vertebrate animals with *dipterous* larvae, which for at least a period feed upon the living or dead tissues or, in the case of intestinal myiasis, on the host's ingested food.

10.1 TYPES OF MYIASIS

Different terms are used to describe myiasis which affects different parts of the body – for example, *cutaneous*, dermal or subdermal myiasis; *urogenital* myiasis; *ophthalmic* or ocular myiasis; *nasopharyngeal* myiasis; and *intestinal*, gastrointestinal or enteric myiasis. When larvae burrow just under the surface layers of the skin this is sometimes called *creeping eruption* or creeping myiasis; when boil-like lesions are produced the term *furuncular* myiasis may be used, and when wounds become infested this is often referred to as *traumatic* myiasis.

Myiasis may be obligatory or facultative. In *obligatory* myiasis it is essential for the fly *maggots* (larvae) to live on a live host for at least a certain part of their life. For example, larvae of *Cordylobia anthropophaga*, *Cochliomyia hominivorax*, *Chrysomya bezziana*, *Dermatobia hominis* and *Wohlfahrtia magnifica* are obligatory parasites of humans and other vertebrates. In contrast, in *facultative* myiasis larvae are normally free-living, often attacking carcasses, but under certain conditions they may infect living hosts. Several types of fly, including species of *Calliphora*, *Lucilia* (= *Phaenicia*), *Phormia* and *Sarcophaga*, which normally breed in meat or carrion, may cause facultative cutaneous myiasis in people by infecting festering sores and wounds.

There is no obligatory intestinal myiasis of humans. When people have maggots in their intestinal tract this is most likely due to accidental swallowing of eggs or larvae on food. Although the maggots may be able to survive for some time in the intestine, there is no species of fly which is specially adapted to cause intestinal myiasis in humans. In contrast, obligatory intestinal myiasis occurs in animals. The presence of larvae in the human intestine may nevertheless cause considerable discomfort, abdominal pain and diarrhoea, which may be accompanied by discharge of blood and vomiting. Living larvae may be passed in excreta or vomit. Occasionally facultative urogenital myiasis occurs in humans; this usually involves larvae of *Musca* or *Fannia* species. Ovipositing flies attracted to unhygienic discharges lay their eggs near genital orifices, and on hatching the minute larvae enter the genital orifice and pass up the urogenital tract. Considerable pain may be caused by larvae obstructing these passages, and mucus, blood and eventually larvae may be discharged during urination.

When larvae occur in wounds, sores and dermal or subdermal tissues, their removal under aseptic conditions is usually a relatively simple procedure. When, however, they are more deeply imbedded in the underlying tissues, or when they have penetrated the mucous membranes, eyes,

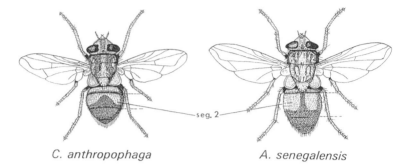

C. anthropophaga   *A. senegalensis*

Figure 10.1 Adults of the tumbu fly (*Cordylobia anthropophaga*) and the Congo floor-maggot fly (*Auchmeromyia senegalensis*) showing the longer second abdominal segment in *A. senegalensis*.

frontal sinuses or cavities, their removal is more difficult and surgery may be needed. Major and irreversible damage may have been done by the larvae.

The biologies and medical importance of the principal types of flies causing facultative and obligatory myiasis in humans are outlined below.

10.2   CLASSIFICATION

Flies described in this chapter are in three families: the Calliphoridae, Sarcophagidae and Oestridae. The family Calliphoridae can be conveniently divided into two groups. One group contains the ***non-metallic*** flies such as the Congo floor-maggot fly (*Auchmeromyia senegalensis*) and the tumbu fly (*Cordylobia anthropophaga*). The other group contains the ***metallic*** calliphorids, such as the blowflies, comprising the bluebottles (*Calliphora*) and the greenbottles (*Lucilia* (= *Phaenicia*)), the New World screwworms (*Cochliomyia hominivorax*) and the Old World screwworms (*Chrysomya bezziana*).

10.3   CALLIPHORIDAE: NON-METALLIC FLIES

10.3.1 *Cordylobia anthropophaga*
External morphology
Cordylobia anthropophaga, known as the tumbu or mango fly, is found in Africa from Ethiopia in the north through West and East Africa to Natal and the Transvaal in the south. It is not found outside Africa except possibly in south-western Saudi Arabia.

Adults are robust, relatively large flies, 9–12 mm long, dull yellowish to light brown in colour, but with two dark grey poorly defined dorsal longitudinal thoracic stripes (Fig. 10.1). There are four visible abdominal segments which are more or less ***equal*** in length (compare *Auchmeromyia*

Figure 10.2 Final-instar larva of the tumbu fly (*Cordylobia anthropophaga*).

senegalensis in which the second abdominal segment is markedly *longer* than the others). The wings are slightly brownish.

Life cycle

Females lay 100–300 eggs in several batches on dry soil and sand in shady places, especially those contaminated with the urine or excreta of humans, rodents, dogs or monkeys. Females also oviposit on underclothes or soiled babies' nappies (diapers) placed on the ground to dry. The white banana-shaped eggs hatch after about 1–3 days. Larvae attach themselves either directly to a suitable host or to washed clothing placed on the ground to dry, and subsequently get transferred to people if the clothing is not ironed before it is worn. On a host, powerful hook-like mouthparts enable a larva to bury itself completely except for its posterior spiracles situated at the tip of the abdomen, which remain in contact with the air. Newly emerged larvae can remain alive for 9–15 days on the ground in the absence of a suitable host.

The minute first-instar larvae are maggot-shaped (like the larvae of house-flies: see Fig. 9.4b), the second-instar larvae are club-shaped, and the third and final instar larvae are rather fat, broadly oval-shaped, yellowish-white maggots, and about 11–15 mm long. They are covered with numerous *spicules* which are often, but not always, grouped into three or more transverse rows per segment (Fig. 10.2). After 8–12 days mature larvae wriggle out of boil-like swellings and fall to the ground where they bury themselves and turn into puparia. Adult flies emerge 8–15 days later and feed on rotting fruit, carrion and faeces. Adults readily enter houses, where they may lay their eggs on mud floors, especially if children have urinated on them.

Medical importance

Larvae of *Cordylobia* cause boil-like (furuncular) swellings on almost any part of the body. Although these swellings may become sore and inflamed, and even quite hard and exude serous fluids, they do not usually contain pus. Generally just one or two larvae are found in a patient; however, more than 60 larvae have been recovered from a person, mostly in separate lesions. The standard method of extracting a larva is to cover the small hole in the swelling with medicinal liquid paraffin. This prevents the larva from breathing through its posterior spiracles with the result that it wriggles a little further out of the swelling to protrude the spiracles. In so doing it lubricates the pocket in the skin, and the larva can then usually be extracted by gently pressing around the swelling.

Infections can be prevented, or at least minimized, by ensuring that clothes, bed linen and towels are not spread on the ground to dry, and also by ironing clothing.

Dogs and rats are commonly infected with tumbu larvae.

10.3.2 *Auchmeromyia senegalensis*
External morphology
Auchmeromyia senegalensis, commonly known as the Congo floor-maggot fly, although not strictly speaking causing myiasis, is described here because adults are often confused with those of *Cordylobia anthropophaga*. The species occurs throughout Africa south of the Sahara and also in the Cape Verde Islands.

Adults are very similar to *C. anthropophaga* but are distinguished by the second abdominal segment being about **twice as long** as any of the others (Fig. 10.1), whereas in the tumbu fly all segments are about equal in length.

Life cycle and medical importance
Some 300 eggs are laid in batches of about 50 on the dry sandy floors of mud huts. They hatch after 1–3 days and the larvae hide in cracks and crevices in the hut floor, especially under beds and sleeping-mats. At night they crawl from these daytime refuges and take blood-meals from people sleeping on the floor. After taking a full blood-meal the now pinkish larvae return to their hiding places. Larvae may feed four or five times a week, but can withstand starvation for long periods in the absence of suitable hosts. There are three larval instars each requiring at least two blood-meals. Under optimum conditions larval development is completed within 3–4 weeks but may be prolonged for up to 3 months if the larvae fail to obtain regular feeds. The fully developed third-instar larvae, unlike those of *Cordylobia anthropophaga*, are not covered with conspicuous spicules (Fig. 10.3). Because larvae cannot climb, people will not be attacked if they sleep on beds raised from the floor. Mature larvae pupate in cracks or directly on the surface of the mud floor of huts. Adults emerge from the puparia after about 9–16 days, and like tumbu flies feed on rotting fruit and faeces.

The Congo floor-maggot fly was formerly common in certain parts of Africa, but due to changes in lifestyle of many people it is becoming increasingly rare, and these days is not much of a problem.

Figure 10.3 Final-instar larva of the Congo floor-maggot fly (*Auchmeromyia senegalensis*).

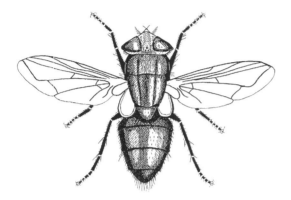

Figure 10.4 Adult of the New World screwworm fly (*Cochliomyia hominivorax*).

10.4 CALLIPHORIDAE: METALLIC FLIES

10.4.1 New World screwworms

The New World screwworm (*Cochliomyia hominivorax*) formerly occured in the southern states of the USA, through Mexico, Central America, the Caribbean islands, into the northern parts of South America. However, following successful eradication programmes, achieved by the release of thousands of sterilized male flies, the screwworm has been eradicated from the USA, Mexico and parts of Central America. An eradication programme is presently continuing in Central America to push the fly's northern limits to Panama. In 1988 a population of the New World screwworm was discovered in Libya, but it was eradicated in 1991.

External morphology

Adult flies of *C. hominivorax* are 8–10 mm long, metallic green to bluish green, and have three distinct dark longitudinal stripes on the dorsal surface of the thorax (Fig. 10.4). The **dorsal bristles** on the thorax, like those of *Chrysomya*, are poorly developed, thus distinguishing screwworm flies from *Lucilia* (= *Phaenicia*) and *Calliphora*, which have well-developed bristles. The thoracic **squama** is a membranous lobe on the posterior border

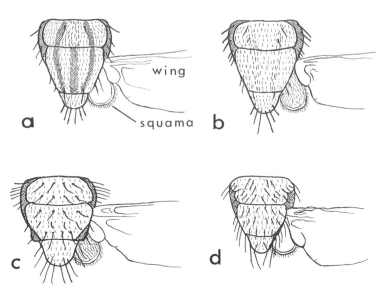

Figure 10.5 Thoraces of adult flies showing the presence or absence of prominent dorsal bristles, and bases of the right wings showing the presence or absence of fine hairs on the squama: (a) *Cochliomyia*, note the three dark thoracic stripes and squama lacking hairs; (b) *Chrysomya*, note the absence of prominent thoracic stripes and the hairy squama; (c) *Calliphora*, note the prominent thoracic bristles and hairy squama; (d) *Lucilia*, note the prominent thoracic bristles and squama lacking hairs.

of the wing near the thorax, and in New World screwworms (Fig. 10.5a) is similar to that of *Lucilia* in being ***devoid*** of hairs.

Life cycle

Females of *C. hominivorax* lay batches of 10–400 eggs on the edges of wounds, scabs, sores, scratches or pimples, also on dried blood clots and on diseased and even healthy mucous membranes such as the nasal passages, eyes, ears, mouth and vagina. In newborn babies eggs may be laid in the umbilicus. Eggs hatch after 11–24 hours and the active larvae burrow deeply into living tissues and feed gregariously. There are three larval instars and the third-instar larvae, which are formed after 2–3 days, are about 15–17 mm long and typically ***maggot***-shaped. They are distinguished from house-fly maggots by the presence of distinct ***bands of spicules*** encircling the anterior margins of all body segments (Fig. 10.6a). Larvae tend to penetrate deeply into tissues, so that infections near the eyes, nose and mouth can cause considerable destruction of these areas, often accompanied by putrid-smelling discharges and ulcerations.

Figure 10.6 Final-instar larvae: (a) *Cochliomyia* (*Chrysomya* larvae are similar) showing bands of spicules on the segments; (b) *Lucilia* (*Calliphora* larvae are almost identical) lacking prominent spicules.

After 4–12 days the larvae reach maturity and wriggle out of the wounds or passages they have excavated by devouring living tissue and drop to the ground, where they bury in the soil and pupate. In warm weather the puparial stage lasts about 7–10 days, but in cooler weather it may be prolonged for many weeks or even months. The life cycle from egg to adult is usually about 2–3 weeks.

10.4.2 Old World screwworms
The genus *Chrysomya* contains many species and is common in the tropics. About 10 species are known to cause myiasis in humans, but only *Chrysomya bezziana* is important because its larvae, like those of *C. hominivorax*, are **obligatory** parasites of living tissues. Larvae of other species of *Chrysomya* are not obligatory parasites and often develop in carrion and decomposing matter. *Chrysomya bezziana* occurs throughout tropical Africa, the Indian subcontinent, in most of South-east Asia to China, and the Philippines to Papua New Guinea; it also has been introduced to several countries on the west coast of the Persian Gulf, including Iraq. Other less important facultative myiasis-producing species such as *Chrysomya albiceps* and *C. megacephala* of the Old World have invaded Central and South America.

External morphology
Adults of *Chrysomya bezziana* are similar to those of *Cochliomyia hominivorax*, but they lack the distinctive longitudinal thoracic stripes (Fig. 10.5b), and the dorsal surface of the thoracic **squama** is covered with fine hairs (Fig. 10.5b). Larvae are very similar to those of *C. hominivorax*, although they differ in usually having 4–6 finger-like processes on the anterior spiracles, not the usual 7–9 found in *C. hominivorax*. In practice separation is usually

easy: the screwworms of the Americas belong to the genus *Cochliomyia* whereas those of the Old World belong to the genus *Chrysomya*.

Life cycle
The life cycle of *Chrysomya bezziana* is very similar to that of *Cochliomyia hominivorax*. About 150–500 eggs are deposited in wounds, open sores, scabs, ulcers, scratches or on mucous membranes, especially those contaminated with discharges. They hatch after 24 hours or less and the newly emerged larvae burrow through the skin to the underlying tissues where they commonly remain congregated together. Larvae complete their development in 5–6 days and then wriggle out of the wounds and drop to the ground, where they bury themselves and pupate. The puparial period lasts about 7–9 days in warm weather, but is prolonged to several weeks or even months during cold weather. The life cycle from egg to adult usually takes about 2–3 weeks.

Adults of both the New and Old World screwworms are frequently found feeding on decomposing corpses, decaying matter, excreta and flowers.

Medical importance of screwworms
Larvae of both *Chrysomya bezziana* and *Cochliomyia hominivorax* are **obligatory** parasites of living tissues and cause human myiasis, which can be very severe, resulting in considerable damage and disfigurement, especially if the face is attacked. When larvae invade natural orifices, such as the nose, mouth, eyes or vagina, they can cause excruciating pain and misery. In one patient suffering from a nasal infection 385 larvae of *C. hominivorax* were removed over a 9-day period! Larvae of both species may eat their way through the palate and as a result impair speech.

Chrysomya bezziana causes more cases of myiasis in people living in India and other parts of Asia than it does in Africa. *Chrysomya* adults often visit faeces, so are potential vectors of several pathogens.

All myiasis cases should be treated immediately because the very rapid larval development can cause permanent damage. Formerly maggots in open wounds or body openings were often removed by irrigating infested areas with 5–15% chloroform mixed with vegetable oil, but because of health and safety regulations, ethanol can be substituted for chloroform. Surgery may be necessary to expose deeply embedded larvae. There are serious problems with ocular myiasis and delicate surgery may be needed to remove the larvae; permanent damage may have been done.

Both screwworm species, but particularly *C. hominivorax*, cause myiasis in cattle, goats, sheep and horses and are responsible for enormous economic losses to the livestock industry. The technique of releasing millions of laboratory-reared sterile male flies into areas infested with screwworms has given good control, and has even resulted in the eradication of *C. hominivorax* from the USA, Mexico, Guatemala, Belize, Honduras, El Salvador, Nicaragua, Costa Rica and some Caribbean islands, as well as from Libya.

10.4.3 The greenbottles (*Lucilia* = *Phaenicia*)
External morphology

There are several species of greenbottles in the genus *Lucilia* and although the genus has a worldwide distribution most species occur in northern temperate regions. Some entomologists use the generic name *Phaenicia* instead of *Lucilia*. Adults are mostly metallic or coppery green, usually a little smaller (about 10 mm long) and a little less bristly than species of *Calliphora* (bluebottles). As in *Calliphora*, prominent **bristles** are present on the dorsal surface of the thorax (Fig. 10.5d), but whereas the **squama** of the wing is hairy dorsally in bluebottles it is without hairs in *Lucilia* (Figure 10.5d). *Lucilia sericata* is the commonest species, occurring in the Americas, Europe, Asia, Africa, Australia and in most other areas of the world. Another common species, *L. cuprina*, occurs mainly in Africa, Asia and Australia.

Life cycle

Female greenbottles normally lay their eggs on meat, fish, carrion and decaying or decomposing carcasses, but occasionally on or near festering and foul-smelling wounds of humans and animals, and on excreta and decaying vegetable matter. Eggs hatch within 8–12 hours. Larvae are typically **maggot**-shaped (Fig. 10.6b). They may bear very small spicules but not conspicuous ones as do larvae of screwworms, which they otherwise tend to resemble. They can also be distinguished from screwworms by the posterior spiracles having a complete **peritreme**; in screwworms it is incomplete.

The larval period lasts about 4–8 days. Mature larvae bury in loose soil and pupate; the puparial period lasts about 6–14 days.

Adult flies frequently visit carrion, excreta, general refuse, decaying material, sores and wounds. They are particularly common around unhygienic places and situations where meat or decaying animals are present. They are nearly always abundant near slaughterhouses and piggeries. They commonly fly into houses, where they are particularly troublesome because of their noisy buzzing flight. The most common species infesting wounds of humans are *Lucilia sericata* and *L. cuprina*.

10.4.4 The bluebottles (*Calliphora*)
External morphology

There are numerous species of flies, known as bluebottles, belonging to the genus *Calliphora*. Although *Calliphora* has a worldwide distribution bluebottles are more common in the northern temperate regions than in the tropical or southern temperate regions. The most important species with regard to myiasis are *Calliphora vicinia* and *C. vomitoria*.

Adults are robust flies, mostly dull metallic-bluish or bluish-black and 8–14 mm long. As in *Lucilia* there are well-developed **bristles** on the thorax (Fig. 10.5c), but the **squama** of the wing is hairy on the dorsal surface (Fig. 10.5c) whereas in *Lucilia* it lacks hairs. The abdomen is rather more shiny than the thorax.

Life cycle
Bluebottle larvae look very similar to those of *Lucilia*, and the life cycle is also very similar to that described for *Lucilia*.

Medical importance of greenbottles and bluebottles
The dirty habit of blowflies (greenbottles and bluebottles) of alighting and feeding on excreta, decaying material and virtually all foods of humans makes them potential vectors of numerous pathogens. However, their medical importance is usually associated with *facultative* myiasis.

Larvae of both *Lucilia* and *Calliphora* have been found in many parts of the world developing in foul-smelling wounds and ulcerations, especially those producing pus. They have also been recorded in hospitals underneath the bandages and dressings of patients, especially when these have become contaminated with blood and pus. Such infections do not usually cause any serious damage or harm since the larvae feed mainly on pus and dead tissues. Removal of maggots of *Lucilia* and *Calliphora* usually presents no problems because they can be picked out of wounds with sterile forceps and antibiotic dressings applied. Very rarely maggots which have infected wounds invade healthy tissues. Occasionally intestinal myiasis is reported. This is usually caused by eating uncooked foods contaminated with larvae of *Lucilia* or *Calliphora*, but usually the larvae are killed within the human alimentary canal and no serious harm is done. As previously emphasized (p. 149) there is no obligatory intestinal myiasis in humans.

Maggot therapy
Australian aborigines and people in northern Myanmar have used maggot therapy for many centuries, and this practice continued in some parts of the world until about the 1950s. However, it seems that during the American Civil War (1861–65) the Confederate army surgeon J. F. Zacharias was the first western physician to introduce maggots to clean wounds. Recently maggot therapy has been rediscovered and doctors, particularly in the USA and Britain but also in some other European countries, are advocating the use of sterile larvae to clean up leg ulcers, pressure sores and infected surgical wounds. Fly maggots most frequently used are those of *Lucilia sericata*. Some patients need considerable persuasion to accept this unusual form of treatment!

S. carnaria   *W. magnifica*

Figure 10.7 Adults of the flesh-flies *Sarcophaga canaria* and *Wohlfahrtia magnifica* showing differences in their abdominal markings.

Control of greenbottles and bluebottles

The principal breeding places of greenbottles and bluebottles are domestic refuse, rubbish tips, dustbins, offal and other wastes of slaughterhouses and meat packing factories. Larvae also occur in foods such as meat and fish left out in the sun to dry. Any methods which reduce the accumulation of these potential breeding sites are welcome. Dustbins and garbage cans should have tight-fitting lids and be emptied once or twice a week. The outside of such bins and both sides of the lid can be sprayed with organophosphate or carbamate insecticides every 7–10 days, and adjacent walls and fences sprayed every 2 weeks. This will deter flies from ovipositing in these sites.

Many of the insecticidal control measures used against house-flies and related flies described in Chapter 9 are applicable to controlling blowflies.

10.5 SARCOPHAGIDAE: FLESH-FLIES

Only species in the genera *Sarcophaga* and *Wohlfahrtia* of the Sarcophagidae are of medical importance. They cause myiasis, and possibly act as mechanical vectors of pathogens. They are sometimes called flesh-flies. They are unusual in that females are *larviparous*, that is they deposit first-instar larvae instead of laying eggs. They have a worldwide distribution.

10.5.1 *Sarcophaga*
External morphology

Sarcophaga species are large and hairy *non-metallic* flies, about 10–15 mm long. They are usually greyish and have three prominent black longitudinal dorsal stripes on the thorax. The abdomen is sometimes distinctly, but other times indistinctly, marked with squarish dark patches on a grey background giving it a *chequer-board* (chess-board) appearance (Fig. 10.7).

Figure 10.8 Larva of *Sarcophaga* showing bands of spicules on the segments and the deep pit in which the posterior spiracles are located (*Wohlfahrtia* larvae are almost identical).

Life cycles and medical importance

Adult *Sarcophaga* do not lay eggs but deposit first-instar larvae, as do tsetse-flies and *Wohlfahrtia* species (i.e. they are **larviparous**). Larvae are deposited in batches of 40–60, usually on decaying carcasses, rotting food and human and animal excreta, but sometimes in wounds. They are primarily scavengers. Larvae are **maggot**-shaped. They are distinguished from larvae of the Calliphoridae by the **posterior spiracles** being situated in a deep pit (Fig. 10.8), and thus difficult to see, and also by having bands of **spicules** on the body. Larvae of *Sarcophaga* are virtually indistinguishable from those of *Wohlfahrtia* species.

Larval development is rapid, lasting in hot weather in the presence of a nutritious food supply only 3–4 days, at the end of which time the larvae bury in the soil and pupate. The puparial stage lasts about 7–12 days.

Although larvae are usually deposited in carrion they very occasionally occur in wounds and cause **facultative** myiasis, but usually causing little damage because they feed mainly on necrotic tissues. They have more commonly been incriminated in accidental intestinal myiasis, causing considerable discomfort and pain before the larvae are passed in the faeces. The most common species is *Sarcophaga cruentata* (= *haemorrhoidalis*), which is widely distributed in the Americas, Europe, Africa and Asia, while *S. canaria* is common in the Palaearctic region. Because adults frequent festering wounds, excreta and decaying animal matter they may be mechanical vectors of various pathogens.

10.5.2 *Wohlfahrtia*

External morphology

Wohlfahrtia species are hairy flies, 8–15 mm long, being about as large as, or a little larger than, bluebottles. They are greyish and like *Sarcophaga* have three distinct black lines on the dorsal surface of the thorax. The dark markings on the abdomen, however, are not in the form of a chess-board pattern as in *Sarcophaga* species, but are usually present as **roundish** lateral spots and triangular-shaped dark markings along the midline (Fig. 10.7). There is, though, considerable variation and sometimes the dark marks are so large as to be more or less confluent, making the abdomen appear mainly black.

Life cycle and medical importance

As with *Sarcophaga* and tsetse-flies, adults of *Wohlfahrtia* deposit **larvae** not eggs. The most important species, *W. magnifica*, causes **obligatory** myiasis in humans and animals (e.g. camels, domestic livestock and dogs) in western Europe, the Middle East, North Africa and central Asia to China. Some 120–170 larvae are deposited, often in several batches, in scratches, wounds, sores and ulcerations. In people the ears, eyes and nose are frequently infested, and this can result in deafness, blindness and even death. The North American species, *W. vigil*, will also deposit its larvae on unbroken skin if it is soft and tender, such as in babies and very young children, who are, therefore, more commonly attacked than adults. The maggots cause **furuncular** myiasis. *Wohlfahrtia* larvae have bands of **spicules**, and the **spiracles** are situated in a deep pit (Fig. 10.8); they closely resemble larvae of *Sarcophaga*. Larval development takes 5–9 days, after which mature larvae drop to the ground, bury themselves in loose soil and pupate. Adults emerge from the puparia after 8–12 days.

Like the larvae of screwworms, *Wohlfahrtia* larvae can burrow deeply into the tissues and cause considerable damage.

10.6 OESTRIDAE: BOT-FLIES

The family Oestridae comprises four subfamilies, three of which (Oestrinae, Gasterophilinae and Hypodermatinae) contain important obligate parasites of domesticated animals. The subfamily Cuterebrinae contains some 83 species in six genera that cause myiasis in rodents, monkeys and livestock, and the human bot-fly (*Dermatobia hominis*). This fly causes **obligatory** myiasis in people and animals, mainly cattle, living in southern Mexico to northern Argentina.

10.6.1 *Dermatobia hominis*
External morphology

Adults of *Dermatobia hominis* are a little larger (12–18 mm) than bluebottles (*Calliphora*) but have a similar dark-blue **metallic-coloured** abdomen, dark bluish-grey thorax and a mainly yellowish head (Fig. 10.9). The mouthparts are vestigial. They are sometimes called the human bot-fly or in the Americas ver macaque or tórsalo.

Life cycle

Dermatobia hominis occurs primarily in lowland forests, being especially common along woodland paths and at the margins of forest and scrub

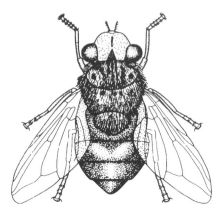

Figure 10.9 Adult of *Dermatobia hominis* (human bot-fly).

areas. These flies have an interesting and remarkable life history. Females glue 6–30 *eggs* to the body of other arthropods, such as day-flying mosquitoes (especially of the genus *Psorophora*), house-flies, stable-flies and occasionally even ticks.

Embryos within the attached eggs mature into first-instar larvae after 4–11 days. They do not hatch, however, until the insect carrying the eggs settles on a human or some other warm-blooded animal to take a blood-meal or, as in the case of house-flies, feed on sweat. Larvae then emerge from the eggs, which remain attached to the insect carrier, and drop on to the host's skin. Here, within 5–10 minutes, the larvae either enter the skin through the bite puncture or penetrate soft unbroken skin and burrow into the subcutaneous tissues. Each larva produces a boil-like (*furuncular*) swelling which has an opening through which the larva breathes. Occasionally two or three larvae may be found within a single swelling.

First-instar larvae are 1–1.5 mm long, more or less cylindrical in shape (Fig. 10.10a), and have the anterior half of the body covered with numerous *spines* of two different sizes. Second-instar larvae are a completely different shape, being enlarged anteriorly but with the posterior half of the body distinctly narrow, giving the appearance of a bottle with a long neck. Relatively large thorn-like *spines* encircle the middle segments (Fig. 10.10b). The third and final instar larvae are about 18–25 mm long, more or less oval and have relatively small *spines* on the anterior segments (Fig. 10.10c). A pair of very distinct *flower-like* spiracles are present anteriorly, and less conspicuous slit-like spiracles are situated in a small concavity of the last abdominal segment.

Larval development is completed in a small pocket excavated in the subdermal layer of the host, and lasts about 4–10 weeks. Mature larvae wriggle out of the skin and drop to the ground where they pupate just

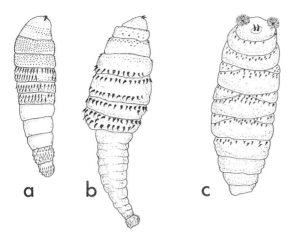

Figure 10.10 Larvae of *Dermatobia hominis* (human bot-fly) showing small and larger spines: (a) first instar; (b) second instar; (c) third and final instar.

under the surface of the soil. Adult flies emerge from the puparia after about 4–11 weeks, but are rarely seen.

Medical importance

Larvae of *Dermatobia hominis* invade the subcutaneous tissues of humans on various parts of the body, including the head, arms, abdomen, buttocks, thighs, scrotum and axillae. They produce boil-like (*furuncular*) swellings which suppurate and this may attract other myiasis-producing flies. They can considerable discomfort and pain. Because of the *long duration* of the larval life (up to 10 weeks or more) infected persons may be encountered in almost any part of the world following air travel.

Because of the numerous spines on the larvae and their shape it is often difficult to remove them by squeezing them out, a procedure that works well with the African tumbu fly; consequently surgical removal of larvae under sterile conditions may be necessary. Frequently a local anaesthetic is needed for this.

10.7   OTHER MYIASIS-PRODUCING FLIES

The black blowfly, *Phormia regina*, found in North America, Europe and northern parts of Asia, breeds mainly in carrion but sometimes causes facultative mysiasis in humans. *Cochliomyia macellaria* is found from North to South America and develops primarily on carrion, but occasionally larvae are found in wounds, and so cause facultative myiasis. It is of only minor medical importance

Several species of flies commonly cause myiasis in livestock. There are, for example, flies infesting donkeys and camels, others infesting sheep and goats (e.g. *Oestrus ovis*), cattle (e.g. *Hypoderma bovis*) and horses and donkeys (e.g. *Gasterophilus haemorrhoidalis*). Occasionally humans become infected with their maggots. Those most likely to suffer are people working with the flies' natural hosts: for example, shepherds looking after sheep may become infected with *O. ovis*.

FURTHER READING

Abram, L. J. and Froimson, A. I. (1987) Myiasis (maggot infection) as a complication of fracture management: a case report and review of the literature. *Orthopedics*, **10**, 625–7.

Arbit, E., Varon, R. E. and Brem, S. S. (1986) Myiatic scalp and skull infection with Diptera *Sarcophaga*: a case report. *Neurosurgery*, **18**, 361–2.

Bunkis, J., Gherini, S. and Walton, R. L. (1985) Maggot therapy revisited. *Western Journal of Medicine*, **142**, 554–6.

Catts, E. P. (1982) Biology of New World bot flies: Cuterebridae. *Annual Review of Entomology*, **27**, 313–38.

Davis, E. and Shuman, C. (1982) Cutaneous myiasis: devils in the flesh. *Hospital Practice*, 1982 (December), 115–23.

Gabaj, M. M., Gusbi, A. M. and Awan, M. A. Q. (1989) First human infestations in Africa with larvae of the American screw-worm, *Cochliomyia hominivorax* Coq. *Annals of Tropical Medicine and Parasitology*, **83**, 553–4.

Graham, O. H. (ed.) (1985) Symposium on eradication of the screw-worm from the United States and Mexico. *Miscellaneous Publications of the Entomological Society of America*, **62**, 1–68.

Greenberg, B. and Kunich, J. C. (2002) *Entomology and the Law: Flies as Forensic Indicators*. Cambridge: Cambridge University Press.

Hall, M. J. R. and Wall, R. (1995) Myiasis of humans and domestic animals. *Advances in Parasitology*, **35**, 257–334.

Kersten, R., Shoukrey, N. M. and Tabbara, K. F. (1986) Orbital myiasis. *Ophthalmology*, **93**, 1228–32.

Krafsur, E. S., Whitten, C. J. and Novy, J. E. (1987) Screw-worm eradication in North and Central America. *Parasitology Today*, **3**, 131–7.

Lane, R. P., Lovell, C. R., Griffiths, W. A. D. and Sonnex, T. S. (1987) Human cutaneous myiasis: a review and report of three cases due to *Dermatobia hominis*. *Clinical and Experimental Dermatology*, **12**, 40–5.

Nunzi, E., Rongioletti, F. and Rebora, A. (1986) Removal of *Dermatobia hominis* larvae. *Archives of Dermatology*, **122**, 140.

Sherman, R. A., Hall, M. J. R. and Thomas, S. (2000) Medicinal maggots: an ancient remedy for some contemporary afflictions. *Annual Review of Entomology*, **45**, 55–61.

Shoaib, K. A. A., McCall, P. J., Goyal, R., Loganathan, S. and Richmond, W. D. (2000) First urological presentation of New World screw worm (*Cochliomyia hominovorax*) myiasis in the United Kingdom: a case report. *British Journal of Urology – International*, **86**, 16–7.

Smith, K. G. V. (1986) *A Manual of Forensic Entomology*. London: British Museum (Natural History); Ithaca, NY: Cornell University Press. pp. 93–137.

Snow, J. W., Siebenaler, A. J. and Newell, F. G. (1981) Annotated bibliography of the screw-worm, *Cochliomyia hominivorax* (Coquerel). *United States Department of Agriculture, Science and Education Administration, Agricultural Reviews and Manuals (Southern Series)*, **14**, 1–32.

Spradbery, J. P. (1991) *A Manual for the Diagnosis of Screw-Worm Fly*. Canberra: Commonwealth Scientific and Industrial Research Organization, Division of Entomology.

Zumpt, F. (1965) *Myiasis in Man and Animals in the Old World: a Textbook for Physicians, Veterinarians and Zoologists*. London: Butterworth.

11

Fleas (Siphonaptera)

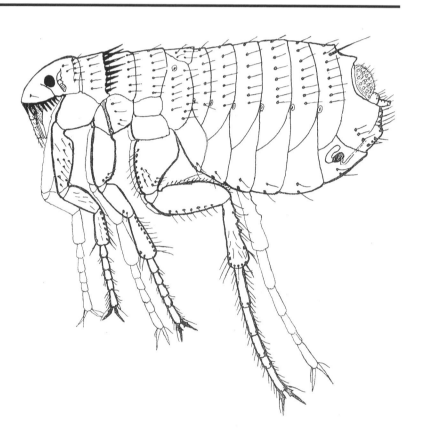

Fleas belong to the order Siphonaptera, and there are some 2500 species and subspecies of fleas in about 230 genera, but only relatively few are important pests of humans. About 94% of known species bite mammals and the remainder are parasitic on birds. Fleas are found throughout most of the world, but many genera and species have a more restricted distribution; for example the genus *Xenopsylla*, which contains important vectors of plague, is confined to the tropics and warmer parts of some temperate countries.

Medically the most important fleas are *Xenopsylla* species, such as *X. cheopis*, which is a vector of plague (*Yersinia pestis*) and flea-borne endemic typhus (*Rickettsia typhi*). Some fleas, such as *Ctenocephalides* species, are intermediate hosts of cestodes (*Dipylidium caninum*, *Hymenolepis diminuta*). Fleas may also be vectors of tularaemia (*Francisella tularensis*), and the chigoe or jigger flea (*Tunga penetrans*) 'burrows' mainly into people's feet.

11.1   EXTERNAL MORPHOLOGY

Adult fleas are relatively small (1–6 mm), more or less oval insects, *compressed* laterally and varying from light to dark brown. Wings are absent, but there are three pairs of powerful and well-developed legs, the hind pair of which are specialized for jumping. The legs, and also much of the body, are covered with bristles and small spines.

The head is approximately triangular in shape, bears a pair of conspicuous black eyes (a few species are eyeless), and short three-segmented more or less club-shaped *antennae* which lie in depressions behind the eyes. The mouthparts point downwards. In some species a row of coarse, well-developed tooth-like spines, collectively known as the *genal comb* or genal ctenidium, is present along the bottom margin of the head (Figs. 11.1, 11.2).

The thorax has three distinct segments: the pro-, meso- and metathorax. The posterior margin of the pronotum (i.e. dorsal part of the prothorax) may bear a row of tooth-like coarse spines forming the *pronotal comb* or pronotal ctenidium (Fig. 11.1). Fleas in some genera lack both pronotal and genal combs and are referred to as *combless* fleas (Fig. 11.2). In some genera fleas have both combs, while in other species the pronotal comb is present and the genal comb absent, but never the reverse (Fig. 11.2). A sternite called the mesopleuron (mesosternum) is located above the middle pair of legs. In several genera, including *Xenopsylla*, which contains important plague vectors, this sternite is clearly divided into two parts by a thick vertical rod-like structure called the *meral rod*, pleural rod, or the mesopleural suture or rod. The presence of this rod, combined with the absence of both genal and pronotal combs, indicates the genus *Xenopsylla* (Fig. 11.2). However, it must be stressed that the presence of a meral rod by itself does not identify fleas as species of *Xenopsylla*, because fleas in several other genera having combs also have a meral rod.

In female fleas the tip of the abdomen is more rounded than in the male and is not upturned, and lying internally in about the position of the sixth to

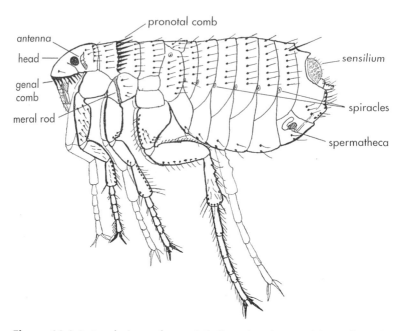

Figure 11.1 Lateral view of an adult flea showing position of combs and the meral rod (pleural rod).

eighth abdominal segments are one or two distinct brownish spermathecae (Fig. 11.1). However, epidemiologically it is not important to separate the sexes because both take blood-meals and can be vectors. A sensory dome-shaped structure having setae, called the sensilium, is present dorsally on apparent segment 8 and aids fleas in detecting vibrations and temperature changes and in host detection.

11.1.1 The alimentary tract of adult fleas

For a better understanding of the role fleas have in the transmission of plague it is necessary to describe the alimentary tract and the method of blood-feeding.

Saliva, which contains anticoagulants such as apyrase, is injected into the host during blood-feeding. Blood is sucked up through the spindle-shaped pharynx and thin oesophagus into the bulbous *proventriculus* (Fig. 11.3). This is provided internally with numerous (250–450) backwardly project-ing spines which when pressed together prevent the regurgitation of the blood-meal into the oesophagus (Fig. 11.3). The proventriculus is important in the mechanism of plague transmission. Finally, the blood-meal enters a relatively large stomach (mid-gut) where it is digested. The distal end of the stomach is connected to the hind-gut, the junction being easily iden-tified by the four Malpighian tubules. The hind-gut is continuous with a

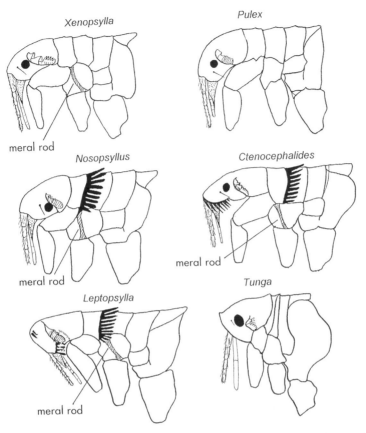

Figure 11.2 Diagrams of the head and thoracic segments of adult fleas showing how the presence or absence of combs and the meral rod can distinguish six medically important genera.

small dilated rectum which has rectal glands which extract water so that the faeces pass out in a dry state.

11.2   LIFE CYCLE

Both sexes of fleas take blood-meals and are therefore equally important as vectors of disease. The following account is a generalized description of the life cycle of fleas that feed on humans or animals, such as dogs, cats and commensal rats. The life cycle of the chigoe (*Tunga penetrans*) is described separately.

A female flea may leave the host to deposit her eggs in debris which accumulates in the host's dwelling place, such as rodent burrows or nests. Species which bite humans or their domestic pets, such as cats and dogs, often lay their eggs in, or near, cracks and crevices on the floor or amongst dust, dirt and debris. Sometimes, however, eggs are laid while the flea is

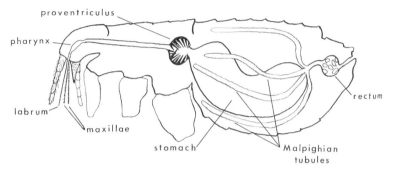

Figure 11.3 Diagrammatic representation of the alimentary canal of an adult flea showing backwardly projecting spines in the proventriculus.

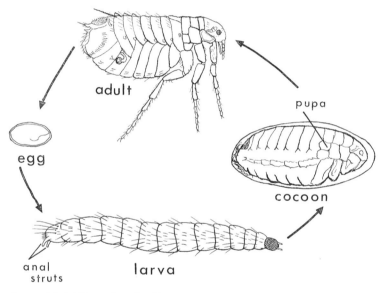

Figure 11.4 Life cycle of a flea.

still on the host and these usually, but not always, fall to the ground. Eggs are very small (0.1–0.5 mm), oval, white or yellowish and lack any pattern. They are thinly coated with a sticky substance which usually results in them becoming covered with dirt and debris. Adults commonly live for 10 days to 6 weeks, but sometimes for 6–12 months or even longer. During her lifetime a female may lay 300–1000 eggs, mostly in batches of 3–25 a day.

Eggs usually hatch after about 5 days but this depends on the species of flea, temperature and humidity. A minute legless larva emerges from the egg (Fig. 11.4). It has a small blackish head with a pair of very small antennae, followed by 13 pale-brown, distinct and more or less similar segments. Each *segment* has a circle of setae near the posterior border. The last segment ends in a pair of finger-like ventral processes termed *anal struts*. The presence of these struts, combined with the setae on the body, distinguish larval fleas from all other types of insects of medical importance.

Larvae are very active. They avoid light and seek shelter in cracks and crevices and amongst debris on floors of houses, or at the bottom of nests and animal burrows. Occasionally, however, larvae are found amongst the fur of animals, and even on people who have unclean habits and dirt-laden clothes, and sometimes in beds. Larvae feed on almost any organic debris including the host's faeces, and partly digested blood evacuated from the alimentary canal of adult fleas; in a few species feeding on expelled blood seems to be a nutritional requirement for larval development. In some species larvae are scavengers and feed on small dead insects or dead adult fleas of their own kind. There are usually three larval instars, but in a few species there are only two instars (e.g. *Tunga penetrans*, p. 176). The larval period is commonly 2–3 weeks, but varies according to species, and may be prolonged to 200 days or more by unfavourable conditions such as limited food supply and low temperatures. Mature larvae are 4–10 mm long. Unlike adult fleas, larvae cannot tolerate extremes in relative humidity and die if humidities are either too low or too high.

At the end of the larval period the larva spins a whitish *cocoon* from silk produced by its salivary glands. Being sticky it soon becomes covered with fine particles of dust, organic debris and sand picked up from the floor of the host's home. Consequently cocoons are difficult to distinguish from their surroundings. About 2–3 days after having spun a cocoon around itself the larva pupates within the cocoon. Adults emerge from the pupa after about 5–14 days, but this period depends on the ambient temperature. The adult flea now requires a stimulus, usually vibrations, such as those caused by the movements of the host within its home, burrow or nest, before it escapes from the cocoon. If, however, animal shelters or houses are vacated, then adult fleas fail to escape from their cocoons until their dwelling places are reoccupied. In some species, carbon dioxide emitted from hosts or a seasonal increase in humidity stimulates emergence. Adults may remain alive in their cocoons for *4–12 months.* This explains why people moving into buildings which have been vacated for many months may suddenly be attacked by large numbers of newly emerged very bloodthirsty fleas seeking their first blood-meal.

The life cycle from egg to adult emergence may be as short as 2–3 weeks for certain species under optimum conditions, but frequently the life cycle is considerably longer, taking many months.

Fleas avoid light and therefore are usually found sheltering amongst the hairs or feathers of animals, or on people under their clothing or even in beds. Given the opportunity many species of fleas feed several times during the day or night on their hosts. During feeding fleas eject faeces composed at first of semidigested blood of the previous meal and then excess blood ingested during the act of feeding. This mixture of partially digested and virtually undigested blood often marks clothing and bed linen of people heavily infected with fleas.

Although most species of fleas have favourite hosts they are not entirely host-specific; for example, cat and dog fleas (*Ctenocephalides felis* and *C. canis*) will readily feed on humans, especially in the absence of their normal hosts. Human fleas (*Pulex irritans*) feed on pigs, and rat fleas in the genus *Xenopsylla* will attack people in the absence of rats. Most fleas will in fact bite other hosts in their immediate vicinity when their normal hosts are absent or scarce. Although feeding on less acceptable hosts keeps fleas alive, their fertility can be reduced by continued feeding on such hosts. Fleas rapidly abandon dead hosts to seek out new ones, behaviour which is of profound epidemiological importance in plague transmission. Fleas can *withstand* both considerable desiccation and prolonged periods of starvation, for example 6 months or more, when no suitable hosts are present. On a host, fleas move either by rapidly crawling or by jumping, but off the host they tend to jump more than crawl in their search for new hosts. Fleas can jump about 20 cm vertically and 30 cm or more horizontally. Such remarkable feats are achieved through a rubber-like protein called resilin, which is very elastic and can become highly compressed; rapid expansion of the compressed state gives the power for jumping.

11.3   MEDICAL IMPORTANCE

11.3.1 Flea nuisance

Although fleas can be important vectors of disease, the most widespread complaint about them concerns the annoyance caused by their bites, which may result in considerable discomfort and irritation. The most common nuisance flea is the cat flea, *Ctenocephalides felis* (four subspecies of *C. felis* are recognized; the true cat flea is *C. felis felis*, but for convenience will be referred to as just *C. felis*). Of lesser importance as a pest is the dog flea, *Ctenocephalides canis*, and morely rarely the human flea, *Pulex irritans*. The cat flea has become the most common flea on dogs.

Fleas frequently bite people on the ankles and legs, but at night a sleeping person is bitten on any part of the body. In many people the bite is felt almost immediately, but irritation usually becomes worse some time after biting. People that become hypersensitive to flea bites can suffer from dermatitis, and inhalation of flea faeces can cause allergies. Children under 10 years tend to experience greater discomfort from flea bites than older people.

Because fleas are difficult to catch this increases the annoyance they cause. People attacked by fleas frequently spend sleepless nights alternately scratching themselves and trying to catch the fleas.

11.3.2 Plague

Bubonic plague, caused by *Yersinia pestis*, is a **zoonosis** being primarily a disease of wild animals, especially rodents, not people. Over 220 rodent species have been shown to harbour plague bacilli. The transmission cycle of plague between wild rodents, such as gerbils, marmots, voles, chipmunks and ground squirrels, is termed **sylvatic**, campestral, rural or endemic plague. Many different species of fleas bite the rodents and maintain plague transmission amongst them. When people such as fur trappers and hunters handle these wild animals there is the risk that they will get bitten by rodent fleas and become infected with plague.

An important form of plague is **urban plague**. This describes the situation when plague circulating among wild rodents has been transmitted to commensal rats (e.g. black rat, *Rattus rattus* and brown (= Norwegian) rat, *R. norvegicus*), and is maintained in the rat population by fleas such as *Xenopsylla cheopis* (Europe, Asia, Africa and the Americas), *X. astia* (Southeast Asia) and *X. brasiliensis* (Africa, South America and India). When rats live in close association with people, such as in rat-infested slums, fleas normally feeding on rats may bite humans. This is most likely to happen when rats are infected with plague and rapidly develop an acute and fatal septicaemia. On their death infected fleas leave the rats and feed on humans. In this way bubonic plague is spread by rat fleas to human populations. The most important vector is *Xenopsylla cheopis* but other species, including *X. astia* and *X. brasiliensis*, more rarely *Nosopsyllus fasciatus* and even more rarely *Leptopsylla aethiopica*, have been incriminated in plague transmission in some areas. However, the last two rodent fleas are reluctant to feed on people and so rarely transmit the disease to humans. In addition to humans becoming infected by the bites of fleas which have previously fed on infected rats, plague can also be spread from person to person by fleas, such as *Xenopsylla* species and *Pulex irritans*, feeding on a plague victim then on another person. This latter method, however, appears to play a minor role in transmission.

It is important to understand the methods by which fleas transmit plague. Plague bacilli sucked up by male and female fleas during blood-feeding are passed to the stomach, where they multiply greatly and extend forwards to invade the **proventriculus**. In some species, especially those in the genus *Xenopsylla*, further multiplication in the proventriculus may result in it becoming partially, or more or less completely, **blocked**. In a partially blocked flea, because the proventriculus is not functioning normally, some of the blood that has been sucked into the stomach, and which is infected with plague bacilli, is regurgitated back into the host (Fig. 11.5). With completely blocked fleas, blood is sucked up from a host with considerable difficulty about as far as the proventriculus, where it mixes with plague bacilli and is then regurgitated back into the host. Blocked fleas soon become starved and repeatedly bite in attempts to get a blood-meal, and

a   b   c

Figure 11.5 Development of *Yersinia pestis* in *Xenopsylla cheopis*: (a) early stage with proventriculus functioning normally; (b) partial blocking with proventriculus failing to form an efficient valve mechanism; (c) completely blocked flea, with blood not being able to enter the stomach (courtesy of Miss M. A. Johnson; and Blackwell Scientific Publications, Oxford, publishers of *Entomology for Students of Medicine* (1962) by R. M. Gordon and M. M. J. Lavoipierre).

therefore are epidemiologically potentially very dangerous. Even if there is no blockage of the alimentary canal, plague transmission can nevertheless occur by direct contamination from the flea's mouthparts.

Another, but less important, method of infection is by the flea's faeces being rubbed into abrasions in the skin or coming into contact with mucous membranes. Plague bacilli can remain infective in flea faeces for as long as 3 years. Occasionally the tonsils become infected with plague bacilli due to crushing infected fleas between the teeth.

In septicaemic plague the bacilli are in the blood and in pneumonic plague they invade the lungs. These form of plague are not transmitted by fleas.

11.3.3 Flea-borne endemic typhus

Flea-borne, endemic or murine typhus is caused by *Rickettsia typhi* (= *mooseri*), the rickettsiae being ingested by the flea with its blood-meal. In the gut the rickettsiae multiply, but unlike plague bacilli they do not cause any blockage of the proventriculus or stomach. Transmission occurs when infected *faeces* are rubbed into abrasions or come into contact with delicate mucous membranes, and also by the release of rickettsiae from crushed fleas. Faeces remain *infective* for many months to a year or more; under laboratory conditions they have remained infective for 4.5–9 years! Murine typhus is essentially a disease of rodents, particularly rats and especially *Rattus rattus* and *R. norvegicus*. It is spread among rats and other rodents by *Xenopsylla* species, especially *X. cheopis*, but also by *Nosopsyllus fasciatus*, *Leptopsylla segnis* and by a few ectoparasites which are not fleas, such as the spined rat louse (*Polyplax spinulosa*) (possibly also *Hoplopleura* species) and possibly by the cosmotropical rat mite (*Ornithonyssus bacoti*). People become infected mainly through *Xenopsylla cheopis*, but occasionally *Nosopsyllus fasciatus*, *Ctenocephalides canis*, *C. felis* and *Pulex irritans* may be involved. *Leptopsylla segnis* does not attack humans, but it is possible that murine typhus is sometimes spread to people by an aerosol of this flea's infective faeces.

Rickettsiae of murine typhus can pass to the flea's ovaries and subsequently to the eggs and larvae, in a form of **transovarial** transmission. But whether this is epidemiologically important remains uncertain.

Recently *Rickettsia felis*, very similar to *R. typhi*, has been isolated in the USA from opposums (*Didelphis* species) and cat fleas which feed on them. It seems there is an urban type of murine typhus involving rats and rat fleas (described above) and a rural type involving opossums and cat fleas (*C. felis*) which can also be transmitted to humans.

11.3.4 Cestodes
Dipylidium caninum is the commonest tapeworm of dogs and cats and occasionally occurs in children. It can be transmitted by fleas to both rodents and humans in the following way. Tapeworm eggs in excreta of domestic pets may be swallowed by larval fleas feeding on the excreta. Larval worms hatching from the ingested eggs penetrate the gut wall of the larval flea and enter the body cavity (coelom). They remain trapped here before passing to the pupa and finally to the adult flea where they encapsulate and become cysticercoids (infective larvae). Animals become infected by licking their coats during grooming and swallowing infected adult fleas. Similarly, young children fondling and kissing dogs and cats can become infected by swallowing cat and dog fleas, or by being licked by dogs which have crushed infected fleas in their mouths, thus liberating the infective cysticercoids.

The rat tapeworms *Hymenolepis diminuta* and *H. nana* have similar life cycles.

11.3.5 Less important diseases
Cat-scratch disease (*Bartonella henselae*) is transmitted among cats by cat fleas. It seems that cats' claws are contaminated with *Bartonella*-infected flea faeces and that transmission to humans is mainly by cat scratches. Tularaemia (*Francisella tularensis*) may occasionally be transmitted to humans by flea bites. Ticks, however, are the main vectors, and it must be stressed that the role of fleas as vectors of these two infections is minimal.

11.4 *TUNGA PENETRANS*

11.4.1 External morphology
Tunga penetrans is found in the tropics and subtropics, from Central and South America, the West Indies across Africa to Madagascar. The flea sometimes occurs in persons in India returning from overseas, mainly Africa, but it is not indigenous to India or elsewhere in Asia. *Tunga penetrans* is sometimes referred to as the chigoe, jigger flea or sand flea. *Tunga penetrans*

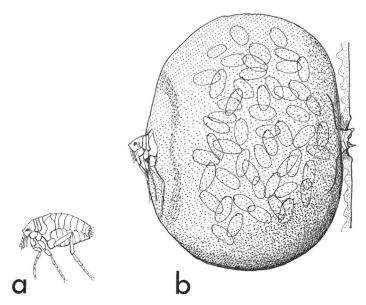

Figure 11.6 Adults of *Tunga penetrans* (chigoe flea): (a) non-gravid female; (b) gravid female with enormously swollen abdomen full of eggs which is embedded in the skin of a host. Note (on right) that the tip of the abdomen projects from the host's skin to the exterior.

does not transmit any disease but is a nuisance because females become imbedded in the skin.

Adults of both sexes are exceedingly small, only about 1 mm long (Fig. 11.6a). They have *neither* genal nor pronotal combs and are easily separated from other fleas of medical importance by their very *compressed* first three (thoracic) segments, and the paucity of spines and bristles on the body.

11.4.2 Life cycle and medical importance

Eggs of *Tunga penetrans* are dropped onto the floor of houses or on the ground outside, and hatch within about 3–4 days. The larvae inhabit dirty and dusty floors or dry sandy soils, especially in areas frequented by hosts of the adult fleas. There are two instars, not three as in most flea species, and under favourable conditions larval development is completed within about 10–14 days; the pupal period lasts about 5–14 days.

Newly emerged adults are very agile and jump and crawl about on the ground until they locate a suitable host, which is usually a person or pig. *Both* sexes feed on blood, but whereas the male soon leaves the host after taking a blood-meal, the female, after being fertilized, 'burrows' into the skin where it is soft, such as between the toes or under toe-nails. Other areas of the foot, including the sole, may also be invaded. In people habitually sitting on the ground, such as beggars or infants, the buttocks may often be infected, and particularly heavy infestations have been recorded from leprosy patients. In heavily infected individuals the arms, especially the elbows, may also be attacked, and occasionally the females 'burrow' into the soft skin around the genital region. Strictly speaking, she does not burrow but the host's skin envelopes her. The entire flea, with the exception of the tip of the abdomen bearing the anus, genital opening and large respiratory spiracles, becomes *completely buried* in the host's skin, and here she continues to feed. The area surrounding the embedded flea becomes very itchy and inflamed, and secondary infections may become established, resulting in ulcerations and accumulation of pus. While the blood-meal is being digested, the abdomen distends to an enormous size and after 8–10 days the flea attains both the shape and size (6 mm) of a small pea (Fig. 11.6b). The abdomen now contains thousands of minute eggs, and over the next 7–14 days about 150–200 eggs are laid, most of which eventually fall to the ground and hatch after about 3–4 days. The life cycle from egg to adult usually takes 4–6 weeks but can be as little as 18 days.

After female fleas die they remain *embedded* in the host. This frequently causes inflammation and may, in addition, result in secondary infections, which if ignored can lead to loss of the toes, tetanus, or even gangrene. Male fleas cause no such trouble as they do not 'burrow' into the skin.

Chigoe fleas are most common in people not wearing shoes, such as children. Up to 100 lesions have been found in the feet of a single child in endemic areas, and there are commonly 40 or more lesions per person. Because the fleas are feeble jumpers, wearing shoes is a simple, but in some communities relatively costly, method of reducing the likelihood of flea infection.

Females embedded in the skin can be removed with fine needles under aseptic conditions, and wounds caused by their extraction sterilized and dressed. They are best removed within the first few days of their becoming established, because when they have greatly distended abdomens, containing numerous eggs, they are difficult to extract without rupturing them, and this increases the risk of infections. Surgical extractions, however, are not easy and it may be better to treat lesions topically with lotions of 0.08% ivermectin or 0.2% metrifonate (trichlorfon) on two consecutive days to kill fleas *in situ*.

Pigs, in addition to humans, are often commonly infected with *Tunga*, and may provide a local reservoir of infection. Other animals such as cats, dogs and rats are also readily attacked.

11.5   CONTROL OF FLEAS

Repellents such as DEET or permethrin-impregnated clothing may afford some personal protection against fleas.

Resistance has been reported in cat fleas, the human flea and *Xenopsylla* species in various parts of the world to one or more of the following categories of insecticides: organochlorines, organophosphates, carbamates, pyrethrins and pyrethroids. Nevertheless insecticides remain the main tool for flea control, although there is increasing reliance on insect growth regulators (IGRs).

11.5.1 Rodent fleas

Fleas in houses can be controlled by applying powders of 5–10% DDT, 1% HCH or 0.5% dieldrin to floors, or with fogs or aerosols containing 2% malathion or 2% fenchlorphos. Insecticidal smoke bombs containing HCH, DDT, permethrin or pirimiphos methyl can also be used to disinfest houses. Insecticidal dusts can be blown into rodent burrows to kill rodent fleas. However, in many parts of the world resistance to DDT, HCH and dieldrin has been reported in *Xenopsylla cheopis* and *Pulex irritans*, so organophosphate insecticides such as 2% diazinon, 2% fenthion, 5% malathion, 2% fenitrothion, 5% iodofenphos or carbamates such as 3–5% carbaryl are used.

Controlling fleas in urban outbreaks of plague or murine typhus usually involves extensive and well-organized insecticidal operations. At the same time as insecticides are applied, rodenticides such as the anticoagulants warfarin and fumarin can be administered to kill rodents. However, if fast-acting 'one-dose' rodenticides such as zinc phosphide, sodium fluoroacetate or strychnine, or the fast-acting anticoagulants such as bromadiolone and chlorophacinone are used, then it is essential to apply these several days after insecticidal applications. Otherwise rodents will be killed before their fleas are killed, and the fleas will then bite other mammals including people, which may result in increasing disease transmission.

Although IGRs appear to have good potential for control of plague fleas they have received relatively little evaluation.

11.5.2 Cat and dog fleas

Cat and dog fleas (*Ctenocephalides felis* and *C. canis*) can be detected by examining the fur on the neck or stomach of the hosts. Proprietary insecticidal powders, sprays or shampoos containing for example 1% HCH, 2–5% malathion, 2–5% carbaryl, 1–2% diazinon or 0.5–1% permethrin can be applied to the animal's fur. Dusts are safer to use than liquids because they are less likely to be absorbed through the animal's skin and cause unpleasant side-reactions.

A simple, but not always very effective, procedure is to use a plastic pet collar impregnated with 20% dichlorvos, 10% propoxur, carbaryl, fenthion or temephos, or an IGR such as methoprene. Flea collars remain insecticidal for 2–3 months, or up to a year with methoprene, but occasionally they may cause skin problems in pets. Alternatively insecticides such as permethrin, fenthion or imidacloprid (a nicotinoid insecticide) can be formulated as a 'spot-on' solution that is applied to the pet's skin in just one small area. The insecticide passes to the animal's blood, so that blood-feeding fleas ingest the insecticide and die. One treatment lasts for about 3–4 weeks.

Insect growth regulators (IGRs) such as methoprene, fenoxycarb or pyriproxyfen are increasingly being applied as topical applications to pets. They provide prolonged ovicidal activity, that is fleas lay non-viable eggs. When IGRs such as cryomazine or lufenuron are administered orally once a month, fleas feeding on the animals produce non-viable eggs for 2–3 weeks, and effective flea control may last for about 4–6 weeks.

However, an important consideration is that most fleas are found away from the host, not on it. For example, it has been said that a typical colony of cat fleas consists of only about 25 adult fleas on the cat, but that on the floor and bedding there may be 500 adult fleas, 500 cocoons and as many as 3000 larvae and 1000 eggs. Clearly, control measures should not be restricted to the cat but applied to beds, kennels, or other places where pets sleep, or spend much of their time. These items should be treated with either insecticidal powders or lightly sprayed with solutions containing 0.5% HCH, 1–2% dichlorvos, 2–5% malathion, 0.5% diazinon, 0.5% chlorpyrifos or one of the pyrethoids, to kill both adult and larval fleas. More recently IGRs such as methoprene, fenoxycarb or pyriproxyfen have been sprayed onto carpets, floors and pets' bedding. Duration of effective control depends on the types of materials sprayed (e.g. earthen or wooden floors, synthetic or woollen carpets), but IGRs generally remain effective longer that conventional insecticides, some giving good control for about 2–4 months. However, flea cocoons are not very susceptible to insecticides, and consequently insecticidal treatments may have to be repeated after 2 months to kill newly emerged adult fleas.

Vacuum cleaning floors, carpets and pets' bedding can also be very effective in removing the immature stages of fleas.

FURTHER READING

Azad, A. F. (1990) Epidemiology of murine typhus. *Annual Review of Entomology*, **35**, 553–69.

Azad, A. F. and Beard, C. B. (1998) Intereactions of rickettsial pathogens with arthropod vectors. *Emerging Infectious Diseases*, **4**, 179–86.

Eisele, M., Heukelbach, J., van Marck, E., Mehlhorn, H., Meckes, O., Franck, S. and Feldmeier, H. (2003) Investigations on the biology, epidemiology, pathology, and control of *Tunga penetrans* in Brazil: I. Natural history in man. *Parasitology Research*, **49**, 557–65.

Hechemy, K. E. and Azad, A. F. (2001) Endemic typhus, and epidemic typhus. In: *The Encyclopedia of Arthropod-Transmitted Infections of Man and Domesticated Animals*, ed. M. W. Service. Wallingford: CABI. pp. 165–9 and 170–4.

Hinkle, N. C., Koehler, P. G. and Patterson, R. S. (1995) Residual effectiveness of insect growth regulators applied to carpet for control of cat flea (Siphonaptera: Pulicidae) larvae. *Journal of Economic Entomology*, **88**, 903–6.

Hinkle, N. C., Rust, M. K. and Reierson, D. A. (1997) Biorational approaches to flea (Siphonaptera: Pulicidae) suppression: present and future. *Journal of Agricultural Entomology*, **14**, 309–21.

Hirst, L. F. (1953) *The Conquest of Plague: a Study of the Evolution of Epidemiology*. Oxford: Clarendon Press.

Pugh, R. E. (1987) Effects on the development of *Dipylidium caninum* and on the host reaction to this parasite in the adult flea (*Ctenocephalides felis felis*). *Parasitological Research*, **73**, 171–7.

Rothschild, M. (1975) Recent advances in our knowledge of the order Siphonaptera. *Annual Review of Entomology*, **20**, 241–59.

Rust, M. K. and Dryden, M. W. (1997) The biology, ecology, and management of the cat flea. *Annual Review of Entomology*, **42**, 451–73.

Schriefer, M. E., Sacci, J. B., Taylor, J. P., Higgins, J. A. and Azad, A. F. (1994). Murine typhus: updated roles of multiple urban components and a second typhuslike rickettsia. *Journal of Medical Entomology*, **31**, 681–5.

Traub, R. and Starcke, H. (eds) (1980) *Fleas*. Proceedings of the International Conference on Fleas, Ashton Wold, Peterborough, UK, 21–25 June 1977. Rotterdam: Balkema.

12

Lice (Anoplura)

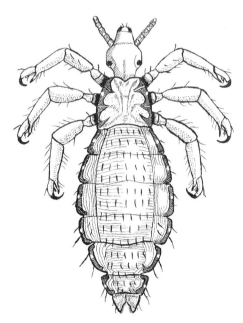

Figure 12.1 Dorsal view of *Pediculus humanus* (body louse). (The head louse, *Pediculus capitis*, looks virtually identical).

There are three types of blood-sucking lice on humans, namely the body louse (*Pediculus humanus*), the head louse (*Pediculus capitis*) and the pubic or crab louse (*Pthirus pubis*). Morphologically the body and head lice are virtually **indistinguishable**. In the laboratory the two can interbreed but very rarely do so in nature. Some regard the head louse as a subspecies of the body louse, but here it is treated as a distinct species. All three species of lice have a more or less worldwide distribution, but they are often more common in temperate areas.

These three species of lice belong to the Anoplura, which has been regarded as either a separate order or a suborder of the order Mallophaga, which contains the chewing lice commonly found on domestic pets and other animals. Here Anoplura is treated as a distinct order.

Body lice are vectors of louse-borne typhus (*Rickettsia prowazekii*), trench fever (*Bartonella quintana*; formerly in the genus *Rochalimaea*) and louse-borne relapsing fever (*Borrelia recurrentis*).

12.1   THE BODY LOUSE (*PEDICULUS HUMANUS*)

12.1.1 External morphology

Adults are small, pale beige or greyish wingless insects, with a soft but rather leathery integument, and are **flattened** dorsoventrally (Fig. 12.1). Males measure about 2–3 mm and females about 3–4 mm. The head bears

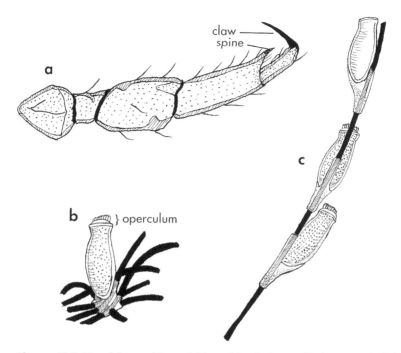

Figure 12.2 Head louse (*P. capitis*) and body louse (*P. humanus*): (a) leg of a body louse showing tarsal claw and tibial spine; (b) unhatched egg of a body louse glued to fibres of clothing; (c) one hatched (upper) and two unhatched (lower) eggs of the head louse cemented to a hair. For convenience these three eggs are shown very close together, but in practice they are rarely this close on a head hair.

a pair of small black eyes and a pair of short five-segmented antennae. The three thoracic segments are fused together. The three pairs of legs are stout and well developed. The short thick tibia has a small thumb-like *spine* on its inner side at the apex, and the short tarsus has a curved *claw* (Fig. 12.2a). Hairs of the host, or clothing, are gripped between this tibial spine and tarsal claw.

Mouthparts of the louse differ from those of most other blood-sucking insects in that they do not constitute a projecting piercing proboscis, but consist of a sucking snout-like projection called the *haustellum*, which is armed on the inner surface with minute teeth which grip the host's skin during feeding. Needle-like stylets are thrust into the skin and saliva injected into the wound to prevent blood from clotting. Blood is sucked up and passes into the stomach for digestion.

The lateral margins of the abdominal segments are sclerotized and much darker than the rest of the segments.

In males there are dark transverse bands on the dorsal surface of the abdominal segments and the tip of the abdomen is rounded, whereas in females it is *bifurcated* and used to grip fibres of clothing during egg laying.

12.1.2 Life cycle

Both sexes take blood-meals and feeding occurs at any time during the day or night. Both adult and immature stages live permanently on humans, clinging mainly to hairs of their clothing and usually only to body hairs during feeding. Female lice glue about 6–10 eggs per day very firmly on hairs of clothing, especially on those along the seams of underclothes, such as vests and pants, but also on shirts and very occasionally on body hairs. The egg, commonly called a *nit* (though some apply this term only to the hatched egg), is oval, white, about 1 mm long, and has a distinct *operculum* (cap) containing numerous small perforations which give the egg the appearance of a minuscule pepper pot (Fig. 12.2b). Intake of air through these holes not only supplies the tissues of the developing embryo with oxygen but aids hatching in the following way. Just prior to hatching, the fully developed louse within its eggshell swallows air which distends the body against the eggshell, thus building up a back pressure causing the head of the louse to be pushed up against the operculum and forcing it off. Female lice live for 2–4 weeks and may lay 150–300 eggs.

The duration of the egg stage is normally about 5–10 days, but eggs on discarded clothing away from the warmth of the body may not hatch until 2–3 weeks; or in cool conditions not at all. Because eggs cannot survive longer than 4 weeks there is little danger of infestation with body lice from clothing not worn for over a month.

Lice have a *hemimetabolous* life cycle. The louse which hatches from the egg is termed a *nymph* and resembles a small adult louse. It takes a blood-meal and passes through three nymphal instars, and after about 7–12 days becomes an adult male or female louse. The duration of the nymphal stages depends on whether clothing is worn continuously. If it is discarded at night and this subjects the nymphs to lower temperatures, this may slow their development. Each louse usually takes several blood-meals a day from its host. The life cycle from egg to adult is about 2–3 weeks.

The body louse is a true ectoparasite of humans. Unfed lice die within about 2–4 days if kept away from humans and without a blood-meal, but blood-fed individuals may live for 5–10 days. Adults probably live for about 30 days. Lice are very sensitive to changes in temperature. They quickly abandon a dead person in response to cooling of the body, and seek out new hosts. They also leave a person with a high temperature, being unable to feed at temperatures above about 40 °C.

A very heavily infested person may have 400–500 lice on their clothing and body. In very exceptional conditions some 30 000 lice have been recorded from a single person! Usually, however, less than 100 lice are found on any one individual, and many have considerably fewer than this.

Body lice are spread by close contact and are especially prevalent under conditions of overcrowding and in situations where people rarely wash or change their clothes. They are therefore commonly found on people in primitive jails, refugee camps and in trenches during wars, and also after disasters such as floods or earthquakes when people are forced to live in very overcrowded, and usually insanitary, conditions. People living in mountainous areas in East Africa, Ethiopia, Sudan, Burundi, Nepal, India and Andean regions of South America, where cold weather necessitates wearing several layers of clothes which may be rarely changed or washed, often have lice. In more developed countries body lice are found mainly on homeless people, and infestations may reach a peak in cold weather when several layers of woollen underclothes are worn.

12.1.3 Medical importance
Pediculosis

The presence of body, head or pubic lice on a person is sometimes referred to as *pediculosis*. The skin of people who habitually harbour large numbers of body lice may become pigmented and tough, a condition known as vagabond's disease, hobo disease or sometimes as *morbus errorum*.

Because lice feed several times a day, saliva is repeatedly injected into people and its toxic effects may cause weariness, irritability or a pessimistic mood: the person feels lousy. Some people develop allergies such as dermatitis or severe itching, or show a type of asthmatic bronchitis. Secondary infections such as impetigo can also result from large numbers of biting lice.

Louse-borne epidemic typhus

Rickettsiae of louse-borne typhus, *Rickettsia prowazekii*, are ingested with the blood-meal taken by both male and female lice, and also their nymphs. They invade the epithelial cells lining the stomach of the louse where they multiply enormously and cause the cells to become greatly distended. About 4 days after the blood-meal the *gut cells* rupture and release the rickettsiae back into the lumen of the insect's intestine. Due to these injuries to the intestinal wall, the blood-meal may seep into the haemocoel of the louse, giving the body a reddish colour. Rickettsiae are passed out with the *faeces* of the louse, and people become infected when these are rubbed or scratched into abrasions, or come into contact with delicate mucous membranes such as the conjunctiva. Infection can also be caused by *inhalation* of the very fine powdered dry faeces. Alternatively, if a louse is crushed, such as by persistent scratching because of the irritation caused by its bites, rickettsiae in the gut are released and may cause infection through abrasions etc. Rickettsiae can remain alive and infective in dried louse *faeces* for about 70 days.

Humans, therefore, become infected with typhus either by the faeces of the louse or by crushing it, not by its bite. An unusual feature of louse-borne epidemic typhus is that it is a disease of the louse as well as of humans. The rupturing of the epithelial cells of the intestine, caused by the multiplication of the rickettsiae, frequently kills the louse after about 8–12 days. This may explain why people suffering from typhus are sometimes found with no, or remarkable few, lice on their bodies or clothing.

People are usually considered to be the reservoir of the disease. Asymptomatic carriers remain infective to body lice for many years. *Recrudescences* as Brill–Zinsser disease, many years after the primary attack, may occur in a person and lead to the spread of epidemic typhus.

Louse-borne epidemic relapsing fever

The causative agent of louse-borne epidemic relapsing fever, *Borrelia recurrentis*, is ingested with the louse's blood-meal from a person suffering from epidemic relapsing fever. Within about 24 hours all spirochaetes have disappeared from the lumen of the gut. Many have been destroyed, but the survivors have passed through the stomach wall to the *haemocoel*, where they multiply greatly, reaching enormous numbers after 10–12 days. The only way a person can be infected is by the louse being crushed and the released spirochaetes entering the body through abrasions or mucous membranes. The habit of some people of crushing lice between the finger-nails, or the even less desirable habit of killing them by cracking them with the teeth, is clearly dangerous if lice are infected with relapsing fever or typhus.

The method of transmission of epidemic relapsing fever must make it very rare for more than one person to be infected by any one infected louse. Hence epidemics of louse-borne relapsing fever will rarely occur unless there are large louse populations.

Trench fever

Trench fever is a relatively uncommon and non-fatal disease which was first recognized during World War I (1914–18) among soldiers in the trenches, and then reappeared in eastern Europe during World War II (1939–45). The disease disappeared again only to reappear later in North America and Europe, occurring mainly in homeless people. It has also been reported from Mexico, North Africa, Burundi and other sub-Saharan countries.

Trench fever is caused by *Bartonella quintana*. The bacteria are ingested by the louse during feeding and become attached to the walls of the gut cells where they multiply; they do not penetrate the cells as do typhus rickettsiae and consequently they are not injurious to the louse. After 5–10 days the *faeces* are infected. Like typhus, the disease is conveyed to humans either by crushing the louse, or by its faeces coming into contact with skin abrasions or mucous membranes. The bacteria persist for many months, possibly even a year, in dried louse faeces, and it is suspected that infection may commonly arise from *inhalation* of the dust-like faeces. The disease may be contracted by those who have no lice but are handling louse-infected clothing contaminated with faeces.

12.1.4 Control

The obvious way to eradicate body lice from a person is by changing and washing the clothing in water hotter than 60 °C, preferably followed by ironing. In epidemic situations, however, such measures may be impractical and immediate reinfestation may occur, so insecticides are usually used for louse control.

Ten per cent DDT dust mixed with an inert carrier (talc) can be blown by a plunger-type duster at the rate of 30–50 g per person, between the body and underclothes. If DDT-resistant lice are present then dusts of 1% malathion, 2% temephos, 1% propoxur or 0.5% permethrin can be used. However, in some parts of the world body lice have also become resistant to malathion and some pyrethroids. Since insecticidal dusts come into close and prolonged contact with people, it is essential that they have very low mammalian toxicities.

Impregnation of clothing with a pyrethroid may provide long-lasting protection against louse infestations, and such treated clothing may remain effective after several washings.

In trials orally administered ivermectin kills body lice, and also head lice, but it is not yet registered for control of human lice.

12.2 THE HEAD LOUSE (*PEDICULUS CAPITIS*)

12.2.1 External morphology

Only very minor morphological differences separate body and head lice. In practice, these differences are not very important because lice found on clothing or on the body are invariably body lice, whereas those on the head are head lice.

12.2.2 Life cycle

The life cycle of the head louse is very similar to that of the body louse except the eggs (*nits*) are not laid on clothes but are cemented to the hairs of the head, usually at their base (Fig. 12.2c), especially behind and above the ears, and at the back of the neck. Usually a single egg is laid on each hair. The distance between the scalp and the furthest egg glued to a hair is often regarded as providing an approximate estimate of the duration of infestation on the basis that a human hair grows about 1 cm per month. However, it appears that eggs may also be laid on long hairs when they are near or touching the scalp, so that unhatched eggs may be some distance from the base of the hair. Only very occasionally are eggs laid on hairs elsewhere on the body. Most individuals harbour only 10–20 head lice, but in very severe infestations the hair may become matted with a mixture of nits, nymphs, adults and exudates from pustules resulting from bites of the lice. In such cases bacterial and fungal infections may become established and an unpleasant crust form on parts of the head, underneath which are masses of head lice. Empty, hatched *eggs* remain firmly cemented to the hairs of the head. A female lays about 6–8 eggs per day, amounting to about 50–150 eggs during her life span, which is a little over 2–4 weeks. Eggs hatch within 5–10 days and the duration of the nymphal stages is about 7–10 days. Away from people head lice die within 2–3 days.

As with body lice, dissemination of head lice is only by close contact, such as children playing together and their heads frequently touching, or when people are crowded together such as in prisons or refugee camps. There is disagreement whether using other people's hats, scarves, combs, hair brushes etc. spreads infections, because it appears that the occasional head lice found on such articles, or on the backs of chairs, are moribund individuals that will not be able to survive and infect someone else. Infestations are often, but not always, greater in women than men, and usually highest in children.

12.2.3 Medical importance

In many areas of the world head lice are a serious public health problem, and in many countries prevalence has been increasing. In some schools in the USA and the UK almost 50% of pupils have head lice. Often there are higher infestation rates in overcrowded homes and where hygiene is poor. There is little evidence that head lice are natural vectors of the diseases transmitted by body lice. For example, typhus epidemics are always associated with body lice, but they may occasionally be minor vectors in some outbreaks of louse-borne relapsing fever.

12.2.4 Control

Regular washing the hair with soap and warm water may reduce the numbers of nymphs and adults, but has no effect on the eggs, which are firmly glued to the bases of hairs. Lice and nits can be removed by a steel or plastic *louse comb* which has very closely set, fine teeth. Alternatively, the head can be shaved. Regular combing with an ordinary comb, although not removing the eggs, may reduce the number of nymphs and adults.

Resistance to DDT, malathion, permethrin, phenothrin and some other pyrethroids has been reported in some parts of the world.

Insecticidal formulations for louse control include dusts, emulsions and lotions. The choice depends on the availability of proprietary brands, preference of patients, degree of residual action required and costs. Although dusts (e.g. 10% DDT and 1% malathion) are efficient, they are not acceptable to most people because they give the head a greyish appearance signalling the person has lice. Preparations such as shampoos which are applied and then washed off after a few minutes are not usually very effective.

In some countries carbaryl and HCH are no longer used against head lice because there is some evidence that they may pose a health risk. Alternative insecticides are emulsions of 1% permethrin, 0.2% phenothrin, 0.5% malathion, 0.3–0.4% bioallethrin or 2% propoxur. Some commercial preparations proclaim that lotions need only remain on the head for 10 minutes or 2 hours, but it is better to leave the insecticidal lotion on the head for about 12 hours, e.g. overnight, before washing it off.

Although some insecticides, such as malathion and permethrin, are reputed to be ovicidal, a *second* treatment after 7–10 days is recommended whatever insecticide is used, because it is difficult to kill all eggs with just a single application. None of the compounds will remove eggs cemented to the hairs, but these can be removed with a louse comb.

Head louse infestations tend to disappear from people who sleep under pyrethroid-impregnated bed-nets used to protect them from mosquito bites (see p. 48).

Experiments using orally administered ivermectin show the drug kills head lice, and also body lice, but as yet ivermectin has not been registered for control of human lice.

12.3 THE PUBIC LOUSE (*PTHIRUS PUBIS*)

12.3.1 External morphology
The pubic louse is generally smaller (1.3–2 mm) than *Pediculus* and is easily distinguished from it. In the pubic louse there is less differentiation between the thorax and abdomen; together they are almost round so that the body is nearly as broad as long. Whereas all three pairs of legs are more or less equal in size in body and head lice, in the pubic louse the front pair, although as long as the other two pairs, is much more slender and has smaller claws. In marked contrast, the middle and hind-legs have massive claws (Fig. 12.3). Presence of a *broad squat* body, *very large* claws on the middle and hind-legs, together with the characteristically more sluggish movements, has resulted in the pubic louse being aptly called the crab louse.

12.3.2 Life cycle
The life cycle of *Pthirus pubis* is similar to that of *Pediculus*. Eggs take about 6–8 days to hatch and the duration of the three nymphal stages is about 10–17 days. Females lay about 3 eggs a day, totalling about 150–200 in their lifetime. Eggs are slightly smaller than those of the body and head louse and are cemented to the coarse hairs of the genital and perianal regions of

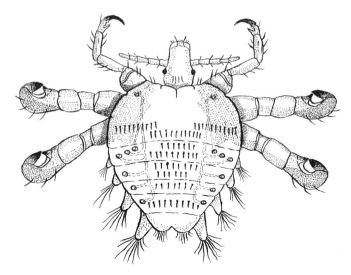

Figure 12.3 Dorsal view of *Pthirus pubis* (pubic louse) showing very large claws on mid- and hind-legs.

the body, and unlike head lice several eggs may be laid on a single hair. Pubic lice may be found on other areas of the body having coarse and not very dense covering of hair, for example, the beard, moustache, eyelashes, underneath the arms and occasionally on the chest. They are very rarely found on the head. Pubic lice are considerably less active than *Pediculus*. The life cycle, from egg laying to formation of the adult, is about 17–25 days.

Infestation with crab lice is usually through sexual intercourse, and characteristically the French call them 'papillons d'amour'. However, it is wrong to suspect that this is the only method. Young children sleeping with parents can catch crab lice from them, and infestations can arise from discarded clothing, infested bedding, or even *rarely* from lavatory seats. Adults survive 2 days or less away from their hosts.

12.3.3 Medical importance

Although in the laboratory pubic lice can transmit louse-borne typhus, there is little evidence that under natural conditions they spread any disease, although it has been suggested that they have been responsible for typhus outbreaks in China. In some individuals, however, severe allergic reactions (*pruritus*) develop to their bites due to the injection of saliva and the deposition around the feeding sites of faeces. Small characteristic bluish spots (*maculae caeruleae*) may appear on infested parts of the body. Infestations of pubic lice are sometimes known as pediculosis pubis or phthiriasis (note that, unlike the generic name of the louse, this is spelt with a 'phth').

12.3.4 Control

Originally control involved shaving pubic hairs from the body, but this method has been replaced by the application of insecticidal emulsions and lotions as used for head louse control.

Basically insecticides used for head louse control can be used against pubic lice. Single applications of 1% permerhrin or 5% malathion should kill eggs, nymphs and adults, but a *second* application 7–10 days later is advisable in case some eggs are not killed. It may be advisable to treat all hairy areas of the body below the neck. Insecticidal shampoos are ineffective for controlling pubic lice. *Aqueous*, not alcoholic, insecticidal solutions should be used otherwise irritation may arise due to sensitivity of the genital region. Insecticide resistance has not yet been reported in pubic lice.

Infestations on the eyelashes can be treated by applying a small amount of petroleum jelly along the closed lashes, twice a day for 10 days. Some physicians have applied 0.25% physostigmine ophthalmic ointment, 1% mercuric oxide ointment for 10–14 days, or fluorescein dye strips to the eyelashes nightly for 3 nights. Only those medically trained should remove eggs from the lashes, which, however, will disappear as lashes fall out and are replaced by others.

FURTHER READING

Bradshaw, W. (1989) How prevalent are headlice? *Parasitology Today*, **5**, 135–6.

Burgess, I. F. (1995) Human lice and their management. *Advances in Parasitology*, **36**, 271–342.

 (1998) Head lice: developing a practical approach. *The Practitioner*, **242**, 126–9.

Buxton, P. A. (1948) *The Louse: an Account of the Lice which Infest Man, their Medical Importance and Control*. 2nd edn. London: Edward Arnold.

Chetwyn, K. N. (1996) An overview of mass disinfestation procedures as a means to prevent epidemic typhus. In: *Proceedings of the 2nd International Conference on Insect Pests in the Urban Environment (ICIPUE)*, ed. K. B. Wildey. The Organising Committee of the ICIPUE. pp. 421–6.

Donnelly, E., Lipkin, J., Clore, E. R. and Altschuler, D. Z. (1991) Pediculosis prevention and control strategies of community health and school nurses. *Journal of Community Health Nursing*, **8**, 85–95.

Gratz, N. G. (1997) *Human Lice: their Prevalence, Control and Resistance to Insecticides. A Review 1985–1997*. CTD/WHOPES/ 97.8. Geneva: World Health Organization.

Maunder, J. W. (1988) Updated community approach to head lice. *Journal of the Royal Society of Health*, **108**, 201–2.

Meinking, T., Burkhart, C. N. and Burkhart, C. G. (1999) Ectoparasitic diseases in dermatology: reassessment of scabies and pediculosis. *Advances in Dermatology*, **15**, 77–108.

Mumcuoglu, K. Y. (1996) Control of lice (Anoplura: Pediculidae) infestations: past and present. *American Entomologist*, **42**, 175–8.

Orkin, M. and Maibach, H. I. (eds.) (1985) *Cutaneous Infestations and Insect Bites.* New York, NY: Marcel Dekker. See Chapters 19–26 on lice.

Service, M. W. (ed.) (2001) *The Encyclopedia of Arthropod-Transmitted Infections of Man and Domesticated Animals.* Wallingford: CABI. See K. E. Hechemy and B. E. Burton, *Bartonella quintana* and *Bartonella henselae*, pp. 70–3; K. E. Hechemy and A. F. Azad, Epidemic typhus, pp. 170–4; and D. A. Warrell, Louse-borne relapsing fever, pp. 295–9.

Sholdt, L. L., Holloway, M. L. and Fronk, W. D. (1979) *The Epidemiology of Human Pediculosis in Ethiopia.* Jacksonville, FL: Navy Disease Vector Ecology and Control Center.

13

Bedbugs (Cimicidae)

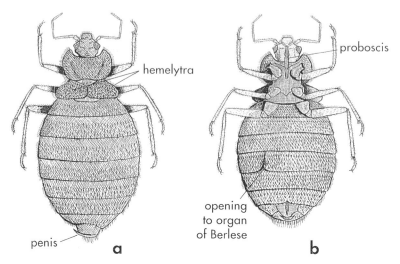

Figure 13.1 Bedbug (*Cimex hemipterus*): (a) dorsal view of adult male; (b) ventral view of adult female.

The family Cimicidae includes the bedbugs, of which two common species feed on humans. *Cimex lectularius* is widely distributed in tropical and non-tropical countries while *C. hemipterus*, commonly called the tropical bedbug, is essentially a species of the Old and New World tropics although it can also occur in warm areas of some non-tropical countries, such as in Florida. It is not easy to separate these two species, but in *C. lectularius* the prothorax is generally 2.5 times as wide as long, whereas in *C. hemipterus* it is only about twice as wide as long. Also, the abdomen is more rounded in *C. lectularius* than in *C. hemipterus*.

A third species, *Leptocimex boueti*, found only in West Africa, bites bats and also people, but is of much less importance than *Cimex* species.

Bedbugs are not considered important vectors, but in addition to constituting a biting nuisance they have been reported as causing iron deficiency in infants.

13.1   EXTERNAL MORPHOLOGY

Adult bedbugs are oval, wingless insects which are *flattened* dorsoventrally (Fig. 13.1). They are about 5–7 mm long and when unfed pale yellow or brown, but after a full blood-meal they become a characteristically darker 'mahogany' brown. The head is short and broad and has a pair of prominent compound eyes, in front of which is a pair of four-segmented antennae. The proboscis is slender, and is normally held closely appressed along the ventral surface of the head and prothorax, but when the bug takes a blood-meal it is swung forward and downwards (Fig. 13.2a). The *prothorax* is much larger than the meso- and metathorax and has distinct wing-like expansions. Two rudimentary and non-functional more-or-less oval wing pads, termed *hemelytra*, overlie the meso- and metathorax. The three pairs of legs are slender but well developed.

The abdomen has 11 segments but only 8 are readily seen. In adult males the tip of the abdomen is slightly more pointed than in the females, while careful examination shows that there is a small well-developed curved penis (Fig. 13.1a). In females there is a small incision ventrally on the left side of the apparent fourth abdominal segment (Fig. 13.1b). This opens into a special pouch (= sinus) called the mesospermalege or the organ of Berlese or Ribaga, which collects and stores sperm. Because both male and female bugs bite people it is not medically very important to distinguish the sexes.

13.2   LIFE CYCLE

Both sexes of bedbug take blood-meals and are equally important as pests. Feeding usually occurs at night on sleeping people, especially just before dawn. If, however, bedbugs are starving they will feed during the day, especially in darkened rooms. Unlike lice, bedbugs do not stay long on people but visit them only to take blood-meals. Very occasionally in temperate climates they remain on vagrants who rarely change their clothes. In the absence of people bedbugs will feed on a variety of hosts, including rabbits, rats, mice, bats, poultry and other birds. During the day adults and nymphs are *inactive* and hide in dark and dry places, such as cracks and crevices in furniture, walls, ceilings or floorboards, underneath seams of wallpaper and between mattresses and beds. At night adults and nymphs crawl from their daytime resting places to feed on sleeping people, after which they return to their resting sites to digest their blood-meals. Bedbugs are gregarious and are frequently found in large numbers. They can move quite rapidly when disturbed.

Females lay about 2–3 eggs a day in cracks and crevices of buildings and furniture, but egg laying ceases at 13 °C or lower. Eggs are about 1 mm long, pearly or yellowish-white, covered with a very fine and delicate mosaic pattern, and characteristically slightly *curved* anteriorly (Fig. 13.2b). Females live several weeks to many months, and during this time may lay 150–200 eggs.

Eggs usually hatch after about 8–11 days, but within less than a week if temperatures are about 27 °C. If, however, temperatures in houses are low, hatching may be delayed for several weeks, and unhatched *eggs* can survive for 3 months. During hatching the small *operculum* (cap) is pushed from the anterior end of the egg, but often remains partially attached. Empty eggshells usually remain cemented in place after hatching. Newly hatched bedbugs (*nymphs*) are very pale yellow and resemble adults, but are much smaller (Fig. 13.2c). The life cycle is *hemimetabolous* and there are five nymphal instars, each of which takes one or more blood-meals.

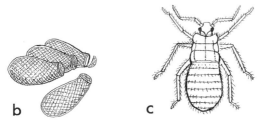

Figure 13.2 Bedbugs: (a) diagram of an adult with proboscis swung forward for blood-feeding; (b) one hatched and three unhatched eggs; (c) first-instar nymph.

The nymphal period lasts from 2–7 weeks, but is greatly extended in cool conditions or if regular blood-feeding is prevented by lack of hosts. The life cycle, from egg to adult (Fig. 13.3), can be as short as 3 weeks if temperatures are high and food plentiful, but is more usually 6–10 weeks. In the laboratory *adults* can live up to 4 years, and survive more than a year without blood-feeding, but survival is dependent on temperature and humidity.

The method of mating in bedbugs is unique among insects. The penis penetrates the integument and enters the mesospermalege (organ of Ribaga or organ of Berlese) situated on the ventral surface of the female abdomen. Sperm introduced into this 'copulatory pouch' (= sinus) pass into the haemocoel. Finally sperm ascend the oviducts and fertlize the eggs.

Bedbug *infestations* can be detected by the presence of live bugs, cast-off nymphal skins, and hatched and unhatched eggs, all of which may be found in cracks and crevices. In addition, small dark brown or black marks may be visible on bed sheets, walls and wallpaper: these are the bedbug's excreta and consist mainly of excess blood ingested during feeding. Houses with large bedbug infestations may have a characteristic sweet and rather sickly smell, but in practice this may not be apparent because the weak odour can be masked by stronger insanitary smells.

Because bedbugs lack wings they do not disperse far although occasionally they crawl from one building to another. Bedbugs are usually spread to new houses by being introduced with furniture and bedding, or more

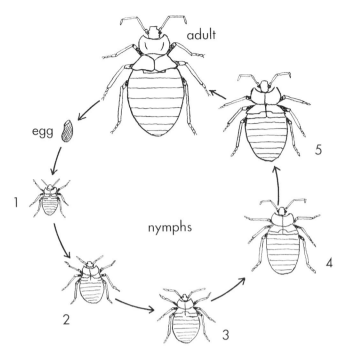

Figure 13.3 Life cycle of bedbugs.

rarely with clothing and hand baggage. Buying secondhand furniture can result in the introduction of bedbugs into houses.

13.3   MEDICAL IMPORTANCE

Although hepatitis B virus and a few other pathogens have been recorded in bedbugs there is no evidence that they can transmit any infections to people. They are therefore not considered to be vectors.

In areas with dilapidated buildings and poor standards of hygiene, bedbug infestations can cause considerable distress. Reaction to their bites is variable. Some people show little or no reaction whereas others may suffer severe reactions and have sleepless nights. Repeated feedings of large numbers of bedbugs have been reported as causing anaemia in infants.

13.4   CONTROL

Insect repellents and pyrethroid-impregnated bed-nets may afford considerable personal protection against bedbugs.

Floors and walls of infested houses together with as much furniture as possible can be sprayed with 5% DDT emulsion. If bedbugs are resistant to DDT then other insecticides such as 1–2% malathion, 0.5% diazinon, 1% carbaryl or 2% propoxur can be used. The addition of 0.1–0.2% pyrethrins or synthetic pyrethroids to sprays is useful because it helps flush out bedbugs from their hiding places, which increases their contact with the insecticide. Pyrethroids, such as 0.5% permethrin, 0.005% deltamethrin or 0.005% lambda-cyhalothrin, can also be used to kill bedbugs.

Mattresses and wooden slats across beds can be dusted with insecticidal powders or lightly sprayed with insecticides (not diazinon), but must be aired afterwards to allow them to dry out completely before being re-used. Bedclothes and infants' mattresses should *not* be treated with insecticides.

Commercially available small insecticidal smoke generators containing, for example, DDT, HCH, permethrin or pirimiphos methyl, which burn for up to 15 minutes, can be used to fumigate infested premises.

FURTHER READING

Anon. (1973) *The Bedbug*. 8th edn. Economic Series 5. London: British Museum (Natural History).

Johnson, C. G. (1941) The ecology of the bedbug, *Cimex lectularius* L., in Britain. *Journal of Hygiene, Cambridge*, **41**, 345–461.

King, F. (1990) Mind the bugs don't bite. *New Scientist*, 27 January, 51–4.

Mayans, M. V., Hall, A. J., Inskip, H. M., Lindsay, S. W., Chotard, J., Mendy, M. and Whittle, H. C. (1994) Do bedbugs transmit hepatitis B? *Lancet*, **343**, 761–3.

Ryckman, R. E., Bentley, D. G. and Archbold, E. F. (1981) The Cimicidae of the Americas and Oceanic Islands, a checklist and bibliography. *Bulletin of the Society of Vector Ecologists*, **6**, 93–142.

Usinger, R. L. (1966) *Monograph of Cimicidae* (*Hemiptera–Heteroptera*). Thomas Say Foundation, volume 7. Maryland: Entomological Society of America.

Venkatachalam, P. S. and Belavady, B. (1962) Loss of haemoglobin iron due to excessive biting by bed bugs: a possible aetiological factor in the iron deficiency anaemia of infants and children. *Transactions of the Royal Society of Tropical Medicine and Hygiene*, **56**, 218–21.

Weidhaas, D. E. and Keiding, J. (1982) Bed bugs. Mimeographed document WHO/VBC/82.857. Geneva: World Health Organization.

14

Triatomine bugs (Triatominae)

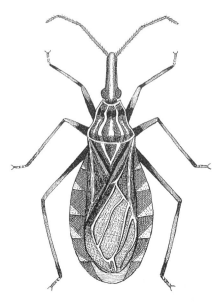

Figure 14.1 Adult *Rhodnius* species as an example of a triatomine bug.

Blood-sucking bugs in the family Reduviidae belong to the subfamily Triatominae and comprise more than 130 species in 15–17 genera (the number depending on whose authority is accepted). Principal species of medical importance are *Triatoma infestans*, *T. dimidiata*, *T. brasiliensis*, *Rhodnius prolixus* and *Panstrongylus megistus*, all of which spread Chagas disease (*Trypanosoma cruzi*) in Central and South America. Some can also transmit *Trypanosoma rangeli*, a non-pathogenic organism.

Most Triatominae occur in the Americas, from the Great Lakes of the USA to southern Argentina, but 13 species are found in the Old World tropics. All medically important species are, however, confined to the southern USA, Central and South America. Triatomines are commonly called kissing-bugs, cone-nose bugs, vinchucas or barbeiros.

14.1 EXTERNAL MORPHOLOGY

Triatominae vary in size from 5 to 45 mm in length, but most are 20–30 mm long. They are easily recognized by their long *snout-like* head which bears a pair of prominent dark-coloured eyes, in front of which is a pair of *laterally* situated, long and thin four-segmented *antennae* (Fig. 14.1). The proboscis, sometimes called the rostrum, is relatively thin and straight and, as in bedbugs, lies closely appressed to the ventral surface of the head (Fig. 14.2a). However, when the Triatominae take a blood-meal the proboscis is swung forward and downwards (Fig. 14.2b). Bugs in other subfamilies of the Reduviidae prey on small insects and the proboscis is

Figure 14.2 Lateral view of the heads of a *Triatoma* species: (a) proboscis closely appressed to ventral side of the head; (b) proboscis swung forward in a blood-feeding position.

Figure 14.3 Fore-wing (hemelytron) of a triatomine bug showing thickened basal part and more membranous distal part.

usually thicker and more robust than in the Triatominae, and is sometimes distinctly curved. Another difference is that the antennae arise from the dorsal surface of the head and not from its sides as in the blood-sucking Triatominae.

The dorsal part of the first thoracic segment of the Triatominae consists of a very conspicuous triangular **pronotum**. The meso- and metathorax are completely hidden dorsally by the folded fore-wings, which are called *hemelytra*. The basal part of each hemelytron is thickened and relatively hard, whereas the more distal part is membranous (Fig. 14.3). The hindwings are entirely membranous but when the bug is not flying they remain hidden underneath the hemelytra. The thorax has three pairs of relatively long and slender legs which end in paired small claws.

The 11-segmented abdomen is more or less oval in shape but is mostly hidden by the wings, except for the lateral margins which are bent upwards slightly and are visible dorsally.

Triatominae are frequently a brown–black colour, but some species are more colourful, having contrasting yellow, orange, pink or red markings, usually as bands on the pronotum, basal part of the fore-wings, legs or margins of the abdomen.

Figure 14.4 Two hatched and two unhatched triatomine eggs.

14.2 LIFE CYCLE

Eggs are about 1.5–2.5 mm long, oval in shape, but have a slight constriction before the *operculum* (cap) (Fig. 14.4). They have a smooth or ornate shell which is pearly white, pink or yellowish depending on the species. Eggs are deposited in or near the habitation of their hosts, such as in cracks and crevices in walls, floors, ceilings and furniture of houses, especially dilapidated mud-walled and thatched-roofed houses in rural areas, or slums at the edge of towns. Adults of some species lay their eggs in rodent burrows and other shelters used by mammalian hosts upon which they feed. Species feeding on birds deposit their eggs in birds' nests or on leaves of trees. Some species lay eggs in small batches which may be glued to the substrate; others lay them more or less singly, either free or cemented to the substrate. Typically females lay 1–2 eggs a day. The total number laid varies from 50 to 1000, depending on the species, their longevity and the number of blood-meals they take, but it is usually 200–300.

The life cycle is *hemimetabolous*. Small pale *nymphs*, which resemble adults but lack wings, may hatch from eggs after only 10–15 days, but the incubation period may extend to 30 or 60 days. Newly emerged nymphs usually remain hidden in cracks and crevices for 2–3 days before they seek out blood-meals. There are five nymphal instars, each requiring at least one blood-meal before it changes into the succeeding instar. Rudimentary wing pads are visible in the fourth and fifth nymphal stages, but only adults have fully developed functional wings. *Young nymphs* can ingest as much as 6–12 times their own weight of blood, and as a result their abdomens may become so greatly distended that they resemble blood-red balloons. Successive instars take relatively less blood, so that the fifth and last nymphal stage takes about 3–5 times its own weight of blood, while adults ingest 2–4 times their weight of blood. In some species adults can ingest 300–400 mg of blood at each meal and feed every 4–9 days! Sometimes hungry nymphs and adult bugs pierce the swollen abdomens of freshly engorged nymphs and take a blood-meal from them without apparently causing harm.

Nymphs and adults of *both* sexes feed at night on their hosts, and *feeding* is a lengthy process often lasting 10–25 minutes. People covered with blankets are bitten on any exposed parts of the body, but especially on the nose and around the eyes and mouth. Biting is usually relatively painless and does not awaken people, but some species cause considerable discomfort and there may be prolonged after-effects. Many Triatominae defecate during or soon after feeding and this behaviour is very important in the transmission of Chagas disease. The presence of bugs in houses is often characterized by streaks of black and white faeces on walls and furniture.

Because of the relatively long time required, even under optimum laboratory conditions, to digest their large blood-meals, the *life cycle* from egg to adult can take 3–4 months but more usually 6–10 months. With large triatomines the life cycle may last a year, or even 2 years. In the absence of hosts, older nymphs and adults can survive 4–6 months of starvation.

Triatomine bugs inhabit both forests and drier areas of the Americas. Many species feed on a variety of wild animals, such as armadillos, opossums, rats, mice, marsupials, ground squirrels, skunks, iguanas, bats, and also birds. Adults and nymphs are usually found in the burrows or nests of these animals. In addition to these sylvatic species, some bugs have become highly domesticated and feed on donkeys, cattle, goats, horses, pigs, cats, dogs and especially chickens, which in some areas appear to be particularly important hosts, and of course humans. These domestic species often live in man-made shelters including houses, especially primitive ones made of wood, mud and thatch. Sylvatic species sometimes move into houses as the forest is cut down and people occupy previously uninhabited areas. Some species are partly sylvatic and partly domestic in their feeding and resting habits.

If hosts vacate their shelters or homes, hungry nymphs crawl from their hiding places to seek new hosts, whereas adults, which are strong fliers, fly out to find new hosts and shelters. Some species are attracted into houses by lights.

14.3   MEDICAL IMPORTANCE

14.3.1 Chagas disease

In rural areas of the Americas there may be hundreds of triatomine bugs in a house and such infestations can be very stressful to the occupiers, who will receive many bites during the night. Typically blood loss can exceed 2 ml per person per night, so it is not surprising that large bug populations can contribute to *anaemia*.

However, medically the main importance of the Triatominae is that they transmit *Trypanosoma cruzi*, the causative agent of Chagas disease, sometimes referred to as South American trypanosomiasis. *Trypomastigotes* ingested with a blood-meal undergo their entire development within the *gut* of the bug. They develop into *epimastigotes* and greatly multiply and after 7–15 days become infective metacyclic trypomastigotes in the lumen of the hind-gut. *Blood-feeding* commonly lasts 10–25 minutes or longer, and during this time, or soon afterwards, many species of bugs excrete liquid or semiliquid faeces which may be contaminated with the meta-cyclic forms of *T. cruzi* derived from a previous blood-meal. People become infected when *excreta* is scratched either into skin abrasions or in the site of the bug's bite, or when it gets rubbed into the eyes or other mucous membranes. If the bug's bite produces local irritation causing the person to scratch, this facilitates infection. *Transmission* is not by the bite of the insect, only through its faeces.

Probably all Triatominae of the Western hemisphere can transmit Chagas disease, and about 70 species have been recorded naturally infected; but some strains of *T. cruzi* are unable to develop in the gut of some species. In practice only about a dozen species living in very close association with people, and therefore regularly feeding on humans, are important vec-tors. Of these the principal ones are *Triatoma infestans* (South America), *Panstrongylus megistus* (mainly Brazil), *Rhodnius prolixus* (mainly Colom-bia and Venezuela), *T. brasiliensis* (eastern Brazil) and *T. dimidiata* (Mexico to Ecuador and Peru). The efficiency of a vector will depend on its speed of feeding and whether or not it defecates on a person during feeding.

Chagas disease is a *zoonosis*. *Trypanosoma cruzi* is essentially a parasite of wild animals, such as opossums (*Didelphis* species), armadillos (*Dasypus* species), many species of wild and urban rats and mice, squirrels, carni-vores, monkeys and possibly bats, all of which may serve as *reservoir hosts*. The *bug* itself is also a reservoir of infection, but in some areas humans are considered to be the principal one. Apart from acquiring infections with *T. cruzi* through faeces, some animals become infected by eating the bugs, or by eating infected animals, such as carnivores eating rodents infected with *T. cruzi*. Rarely people can also become infected by eating infected meat (e.g. inadequately cooked opossums) or food contaminated with excrement of infected bugs.

Infection rates of Triatominae are often exceptionally high. For example, it is not uncommon to find infection rates of about 25% or even 40% or more. Even higher infection rates (78%) have been found in *Triatoma protracta* in California, but because this species very rarely bites people it is not considered a vector to humans.

14.3.2 *Trypanosoma rangeli*

Another trypanosome, *Trypanosoma rangeli*, occurring from Mexico to Brazil and apparently non-pathogenic in humans, is also transmitted by triatomines, especially by *Rhodnius prolixus*. In the vector the trypanosomes undergo dual development: some of the metacyclic infective forms migrate to the hind-gut while others penetrate the gut wall, pass across into the haemocoel and then migrate to the salivary glands. Humans are mainly *infected* by the bug's bite and only rarely by its faeces.

14.4 CONTROL
Control of Chagas disease is based mainly on spraying the interior surfaces of walls and roofs of houses with residual insecticides. Formerly 500 mg/m² of HCH was used, but this has been replaced by the pyrethroids, particularly deltamethrin (25 mg/m²), cyfluthrin (50 mg/m²) and lambda-cyhalothrin (30 mg/m²), and to a lesser extent by cypermethrin (125 mg/m²). A single spraying should eliminate populations of domestic bugs, and sometimes also peridomestic ones. Careful surveillance detects whether there is reinfestation or foci of bugs that need further spraying. Resisistance to several insecticides including some pyrethrioids has been recorded, but presently this does not hinder control operations.

Formulations of paints based on latex or polyvinyl acetate containing insecticides, such as malathion or fenitrothion, which are slowly released can be painted or sprayed onto most indoor surfaces, including mud walls. They can remain effective for up to 2 years.

Insecticidal smoke bombs which when lit dispense pyrethroid insecticides are commonly used in houses.

The above methods will destroy bugs resting in houses but are less effective against those resting in natural outdoor shelters. These peridomestic populations can sometimes invade houses.

Additional methods of control involve making houses unattractive as resting sites for bugs, for example by plastering walls to cover up cracks in which the bugs might live, and by replacing dilapidated mud and thatched houses with those built of bricks or cement blocks and having corrugated metal roofs. However, because of the high costs of building new houses, rehousing has yet to be carried out on a large scale.

14.4.1 'Southern Cone Initiative'
In 1991 Argentina, Bolivia, Brazil, Chile, Paraguay and Uruguay, joined in 1996 by Peru, proposed a plan of action to eliminate vectors, especially *T. infestans*, mainly by vector control methods such as insecticidal spraying of houses.

In 1997 Uruguay was declared free of Chagas disease transmission, as was Chile in 1999 and Brazil in 2001. Other countries, especially Bolivia and Peru, are moving towards interrupted transmission of Chagas. In 1997 two new initiatives were launched to interrupt transmission in the Central American countries of Belize, Costa Rica, El Salvador, Guatemala, Honduras, Nicaragua and Panama, and also in the Andean countries of Colombia, Venezuela and Ecuador.

FURTHER READING
Barrett, T. V. (1991) Advances in triatomine bug ecology in relation to Chagas' disease. *Advances in Disease Vector Research*, **8**, 143–76.

Brenner, R. R. and Stoka, A. de la M. (eds.) (1988) *Chagas' Disease Vectors. I: Taxonomic, Ecological and Epidemiological Aspects.* Boca Raton, FL: CRC Press.

(eds.) (1988) *Chagas' Disease Vectors. II: Anatomic and Physiological Aspects.* Boca Raton, FL: CRC Press.

(eds.) (1988) *Chagas Disease Vectors. III: Biochemical Aspects and Control.* Boca Raton, FL: CRC Press.

Briceno-Leon, R. (1987) Rural housing for control of Chagas' disease in Venezuela. *Parasitology Today*, **3**, 384–7.

Bryan, R. T., Balderrama, F., Tonn, R. J. and Dias, J. C. P. (1994) Community participation in vector control: lessons from Chagas' disease. *American Journal of Tropical Medicine and Hygiene*, **50** (suppl.), 61–71.

Carcavallo, R. U., Galíndez-Girón, I. G., Jurberg, J. and Lent, H. (eds.) (1999) *Atlas of Chagas' Disease Vectors in the Americas. Volume 3.* Rio de Janeiro. Fundación Oswaldo Cruz.

Dias, J. C. P. (1988) Controle de vetores da doença de Chagas no Brasil e riscos da reinvasão domiciliar por vetores secundários. *Memórias do Instituto Oswaldo Cruz*, **83** (suppl. I, special issue), 387–91.

Forattini, O. P. (1989) Chagas' disease and human behavior. In: *Demography and Vector-Borne Diseases*, ed. M. W. Service. Boca Raton, FL: CRC Press. pp. 107–20.

Kingman, S. (1991) South America declares war on Chagas' disease. *New Scientist*, 19 October, 16–7.

Lent, H. and Wygodzinsky, P. (1979) Revision of the Triatominae (Hemiptera, Reduviidae), and their significance as vectors of Chagas' disease. *Bulletin of the American Museum of Natural History*, **163**, 123–520.

Schofield, C. J. (1979) The behaviour of Triatominae (Hemiptera: Reduviidae): a review. *Bulletin of Entomological Research*, **69**, 363–79.

(1988) Biosystematics of the Triatominae. In: *Biosystematics of Haematophagous Insects*, ed. M. W. Service. Oxford: Clarendon Press. pp. 280–312.

(1994) *Triatominae: Biology and Control.* Bognor Regis: Eurocommunica.

Schofield, C. J. and Dias, J. C. P. (1998) The Southern Cone Initiative against Chagas disease. *Advances in Parasitology*, **42**, 1–27.

Schofield, C. J. and Dujardin, J. P. (1997) Chagas disease vector control in Central America. *Parasitology Today*, **13**, 141–4.

World Health Organization (2002) Control of Chagas disease. Second report of the WHO expert committee. *World Health Organization Technical Report Series*, **905**, 1–109.

15

Cockroaches (Blattaria)

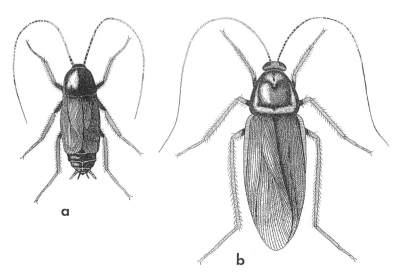

Figure 15.1 Adult cockroaches: (a) *Blatta orientalis* (oriental cockroach); (b) *Periplaneta americana* (American cockroach).

Cockroaches belong to the order Blattaria, and there are about 4000 species. Some 18 species have become serious domestic pests. The most important medically are *Blattella germanica* (the German cockroach), *Blatta orientalis* (the oriental cockroach), *Periplaneta americana* (the American cockroach) and *Supella longipalpa* (the brown-banded cockroach). Cockroaches are sometimes called roaches or steambugs. They have an almost worldwide distribution.

Cockroaches almost certainly aid in the mechanical transmission and harbourage of various pathogenic viruses, bacteria and protozoans.

15.1   EXTERNAL MORPHOLOGY

The more common household pest species are chestnut brown or black, 10–50 mm long, flattened dorsoventrally, and have a smooth, shiny and tough integument. Viewed from above the head appears small and is sometimes almost hidden by the large, rounded *pronotum*. A pair of long and prominent filiform *antennae* arise from the front of the head between the eyes (Fig. 15.1). Cockroach mouthparts are developed for chewing, gnawing and scraping; they are not composed of piercing stylets and therefore cockroaches cannot suck blood. Adults of both sexes have two pairs of wings. In certain household species *wings* in the female are shorter than those of the male, and in female *Blatta orientalis* they are very small and nonfunctional. In cockroaches, the fore-wings, called *tegmina*, are thickened and leathery in texture. They are not used in flight but serve as protective covers for the membranous hind-wings, which when not in use are folded shut, fan-like, over the body. Although the hind-wings can be used for flying cockroaches rarely fly. There are three pairs of well-developed legs covered with prominent small spines and bristles; the five-segmented tarsi end in a pair of claws.

The segmented abdomen is more or less oval in shape but is either completely or partly hidden from view, depending on the species, by the folded overlapping wings. In both sexes a pair of prominent segmented pilose *cerci* arise from the last abdominal segment (Fig. 15.2a), but in some species they are hidden from view by the wings.

Cockroaches are distinguished from beetles (order Coleoptera) by having the *fore-wings* placed over the abdomen in a scissor-like manner. In beetles the fore-wings (elytra) are not crossed over but meet dorsally to form a distinct line down the centre of the abdomen. In addition, the elytra of beetles are generally thicker than the tegmina of cockroaches.

15.2   LIFE CYCLE

Cockroaches like warmth and during the *day* they hide behind radiators and hot-water pipes. In warm countries, where these may be absent, they can be found in almost any dark place, such as cesspits, septic tanks, sewers, refuse tips, dustbins, cupboards, drawers, underneath chairs, tables, sinks, baths and beds, behind refrigerators, cooking stoves, and kitchen crockery. They are common in kitchens, especially if remains of food are left out overnight, and in restaurants, hotels, bakeries, breweries, laundries and aboard ships. They are also found throughout the world in hospitals. Cockroaches are *nocturnal* and are rarely seen during the day unless they are disturbed from their hiding places. They become very active at night, running over floors, tables and other furniture to seek food. When lights are suddenly switched on cockroaches can be both seen and heard scuttling along.

They are *omnivorous* and voracious feeders and any type of food is eaten. They also consume paper, clothes, particularly starched ones, books, hair, wallpaper, dried blood, sputum, excreta, dead insects, and almost any animal or vegetable matter. They have been recorded gnawing finger-nails and toe-nails of babies and sleeping or comatose people, and even infesting the hair of vagrants. Cockroaches habitually disgorge partially digested meals and deposit their excreta on almost anything, including food. They can *live* for 5–10 weeks without water and for many months without food, but this is not an important limiting factor because they very rarely occur in areas where food of some kind is not available. Young nymphs, however, may die within about 7–10 days in the absence of food.

Eggs are laid encased in a brown bean-shaped case or capsule called an *ootheca* (Fig. 15.2b), which can contain 12–50 eggs but typically 18–40. Cockroaches, especially *Blattella germanica*, are often seen running around with an ootheca partly protruding from the tip of the abdomen (Fig. 15.2a).

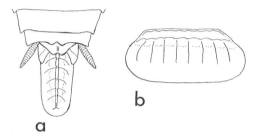

Figure 15.2 (a) Ootheca protruding from the abdomen of *Blatella germanica* (German cockroach); (b) lateral view of a typical cockroach ootheca.

Oothecae are deposited in cracks and crevices in dark and secluded places and in some species are cemented to surfaces, such as the undersides of tables, chairs and beds.

Cockroaches live for many months to 2 years or more, and during this time the female will lay about 4–90 oothecae, the number varying according to species. Cockroaches have a *hemimetabolous* life cycle. Nymphs hatch from eggs after about 1–3 months, the time depending on both temperature and species. Young nymphs are very pale and delicate versions of the adults, whereas the older ones are progressively darker and resemble more the adults. The nymphs are wingless; the wings gradually develop with ensuing nymphal instars, but only adults may have fully developed wings. The number of nymphal stages is variable within and between species. There are commonly 5–7 instars but there may be as many as 13, such as sometimes in *Periplaneta americana*. Duration of the nymphal stage varies according to temperature, abundance of food and species. For example, the *nymphal stage* commonly lasts 2–3 months in *Blattella germanica*, 12–15 in *Blatta orientalis*, and 10–14 in *Periplaneta americana*, but sometimes up to 23 months.

Cockroaches spread very rapidly from infested houses to adjoining ones. They often gain entry by climbing up water pipes and waste pipes. They are also spread as oothecae, nymphs and adults in furniture and other belongings.

15.3   MEDICAL IMPORTANCE

15.3.1 Allergies

Only recently has the importance of cockroach allergies been recognized. About half of asthmatics are allergic to cockroaches, their cast-off skins or excreta, while some 10% of non-allergic people will exhibit cockroach allergies. Symptoms include sneezing, skin reactions, sore eyes and in extreme cases shortness of breath. The allergic rate is second only to that caused by house-dust mites (see Chapter 20).

15.3.2 Infectious agents

Because of their dirty habits of feeding indiscriminately on both excreta and foods, and the practice of excreting and regurgitating their partially digested meals over food, the presence of cockroaches in houses, hotels and hospitals is highly undesirable.

Most of the parasitic infections isolated from cockroaches are spread directly from person to person without the aid of intermediary insects, and so it is very difficult to prove that cockroaches are the prime cause of any disease outbreak. Nevertheless because of their insanitary habits they have been suspected as aiding the transmission of various illnesses. For example, they are known to carry poliomyelitis virus, *Entamoeba histolytica*, *Toxoplasma gondii*, *Escherichia coli*, *Staphylococcus aureus*, *Klebsiella pneumoniae*, *Shigella dysentariae* and *Salmonella* species, including *S. typhi* and *S. typhimurium*. Eggs of the nematode *Enterobius vermicularis*, which is an extremely common worm in humans, can also be carried by cockroaches.

There is little doubt that cockroaches contribute to the spread of several infections, mainly intestinal ones, and possibly they may sometimes be more important as mechanical vectors than house-flies. However, it remains difficult to assess their real importance as vectors because many of the pathogens which cockroaches carry can be transmitted by many other ways.

15.4 CONTROL

Ensuring that neither food nor dirty kitchen utensils are left out overnight will help reduce the number of cockroaches, but if they are present in adjoining or nearby houses, good hygiene in itself will not prevent cockroaches from entering houses.

Insecticidal spraying or dusting selected sites such as cupboards, wardrobes, kitchen furniture, underneath sinks, stoves and refrigerators, and dustbins, is recommended. Resistance to organochlorine and/or organophosphate, carbamate and pyrethroid insecticides is exhibited by many species. In the absence of resistance 0.5–3% sprays or 0.5–5% dusts of organophosphate insecticides such as pirimiphos methyl, fenitrothion, malathion, diazinon or chlorpyrifos, or carbamates such as propoxur, carbaryl or bendiocarb, can be used. Sprays incorporating kerosene (paraffin) may leave unsightly stains on walls, and are of course dangerous near naked flames or cookers.

Pyrethroids (e.g. permethrin, deltamethrin, lambda-cyhalothrin, cyfluthrin) applied as sprays, or more effectively as aerosols, can produce spectacular results in both *flushing out* and killing cockroaches

The residual efficiency of sprays depends on the sprayed surfaces. For example on most painted and shiny surfaces residual activity often lasts only about 1–4 weeks, but for several months on more porous surfaces.

The residual action of insecticidal dusts is less affected by the type of the surface, but dusting is unsightly and so householders may object to it.

Commercial lacquers and varnishes containing residual insecticides such as 1% cypermethrin, 0.5% alpha-cypermethrin or 2% diazinon when painted on walls and other surfaces remain effective in killing cockroaches for several months.

Boric acid powder (borax) still remains a very safe and useful chemical, acting both as a contact insecticide and as a stomach poison.

Organophosphate and carbamate insecticides can be added (1–2%) to baits of food, such as peanut butter, dog food, and maltose, to which glycerol may be added to increase their attractiveness. Such poisonous baits can be placed in areas harbouring large numbers of cockroaches to kill them. Alternatively attractant baits or cockroach pheromones can be placed in simple cardboard or sticky traps to entice cockroaches into them, after which they are either killed or prevented from escaping.

Insect growth regulators (IGRs), such as fenoxycarb, hydrophen and methoprene, have been used to control cockroaches that have developed resistance to a broad spectrum of insecticides, but their use is hindered by high costs and limited availability.

FURTHER READING

Brenner, R. J., Barnes, K. C., Helm, R. M. and Williams, L. W. (1991) Modernized society and allergies to arthropods: risks and challenges to entomologists. *American Entomologist*, **37**, 143–55.

Burgess, N. R. H. and Chetwyn, K. N. (1981) Association of cockroaches with an outbreak of dysentery. *Transactions of the Royal Society of Tropical Medicine and Hygiene*, **75**, 332–3.

Cornwell, P. B. (1968). *The Cockroach. Volume 1: A Laboratory Insect and an Industrial Pest*. London: Hutchinson.

(1976) *The Cockroach. Volume 2: Insecticides and Cockroach Control*. London: Associated Business Programmes.

Guthrie, D. M. and Tindall, A. R. (1968) *The Biology of the Cockroach*. London: Edward Arnold.

Petersen, R. K. D. and Shurdut, B. A. (1999) Human health risk from cockroaches and cockroach management: a risk analysis approach. *American Entomologist*, **45**, 142–8.

Roth, L. M. and Willis, E. R. (1960) The biotic associations of cockroaches. *Smithsonian Miscellaneous Collection*, **141**, 1–470.

Rust, M. K., Owens, J. M. and Reierson, D. A. (eds.) (1995) *Understanding and Controlling the German Cockroach*. New York, NY: Oxford University Press.

Rust, M. K., Reierson, D. A. and Hangsen, K. H. (1991) Control of American cockroaches (Dictyoptera: Blattidae) in sewers. *Journal of Medical Entomology*, **28**, 210–3.

Schal, C. and Hamilton, R. L. (1990) Integrated suppression of synanthropic cockroaches. *Annual Review of Entomology*, **35**, 521–51.

16

Soft ticks (Argasidae)

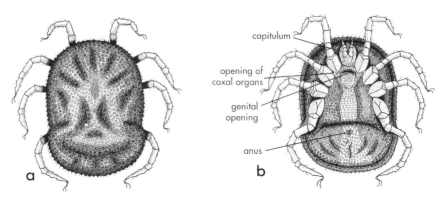

Figure 16.1 Adults of the soft tick *Ornithodoros moubata*: (a) dorsal view; (b) ventral view.

Ticks are not insects because adults have eight legs, not six as in adult insects. Ticks are closely related to mites and spiders. There is some disagreement concerning the classification of ticks and the number of their genera and species. Here ticks are placed in the subclass Acari and order Ixodida. The most important point to remember is that ticks are divided into two main families, namely the Argasidae (soft ticks) and the Ixodidae (hard ticks). A third family, Nuttalliellidae, contains just one species which is of no medical importance. Another important point is that ticks are much *larger* than mites, although their small immature stages may approach the size of some mites, but ticks have a *toothed* hypostome (Fig. 16.2) thus differing from mites.

Soft ticks (Argasidae) have a more-or-less worldwide distribution. There are some 185 species belonging to four genera, but the medically important soft ticks belong to the genus *Ornithodoros*. Species in this genus are found in many areas of the world including the Americas, Africa, Europe and Asia. The most important species is *Ornithodoros moubata*, vector of tick-borne (endemic) relapsing fever, caused by *Borrelia duttonii*. A few other species in the *O. moubata* complex are also of medical importance.

16.1   EXTERNAL MORPHOLOGY

Adult argasid ticks are *flattened* dorsoventrally and oval in outline, but the shape varies according to species. They are 8–13 mm long. The integument is tough, leathery, wrinkled and usually has fine tubercles (mammillae) or granulations. There is *no scutum* (dorsal shield) as is found in ixodid (hard) ticks. The mouthparts, termed the *capitulum*, gnathosoma, or 'false head', are situated ventrally (Fig. 16.1b) and are thus *not visible* dorsally except in their larvae (see p. 217). This character separates adults and nymphal soft ticks, but not larvae, from hard ticks (Ixodidae), which have the capitulum projecting forward and clearly visible dorsally. The four-segmented *palps*

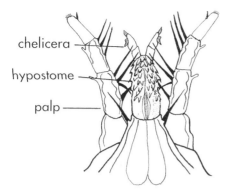

Figure 16.2 Capitulum of an adult *Ornithodoros* species showing leg-like palps and non-denticulate cheliceral sheaths.

are leg-like and the powerful cutting chelicerae have smooth, not denticulate, sheaths and teeth at their tips (Fig. 16.2). The *chelicerae* together with the *hypostome*, which has teeth arranged in several longitudinal rows, penetrate the host during feeding.

There are four pairs of well-developed legs, with each leg terminating in a pair of claws. *Coxal organs* ('glands') open between the bases of the coxae of the first and second pairs of legs (Fig. 16.1b), and are osmoregulatory in function.

Males and females look very similar and are usually difficult to separate, although blood-engorged females can be considerably larger than males because they ingest much greater quantities of blood. However, because *both* sexes feed on blood and can consequently be disease vectors, it is not usually important to distinguish between them.

16.2   INTERNAL ANATOMY

A brief account of the internal anatomy of a tick is necessary to understand the mechanisms of disease transmission.

During feeding, saliva, which usually contains powerful anticoagulants, is secreted by a pair of large grape-like salivary glands and flows down the mouthparts. Blood from the host then passes up the mouthparts, through the narrow oesophagus into the stomach or mid-gut, which is provided with numerous branching diverticula. *Ramifications* of the diverticula enable the adult tick to ingest large volumes of blood (about 6–12 times its own weight), causing great distension of the tick's body.

Argasid ticks have a pair of *coxal organs*, which although sometimes called coxal glands are not actually glandular but filter excess fluid and salts from ingested blood-meals. This fluid passes out through small openings located bilaterally between the bases of the first two pairs of legs. When a soft tick is infected with tick-borne relapsing fever (*Borrelia duttonii*) many

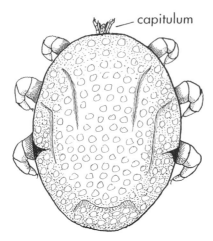

Figure 16.3 Larva of an *Ornithodoros* species, as an example of a typical soft tick showing the capitulum projecting in front of the body.

of the spirochaetes in the haemolymph enter the coxal organs and are passed out through their openings.

Coxal organs are present only in soft ticks, not in hard ticks.

In females of both soft and hard ticks a peculiar structure termed *Gene's organ*, located in front of the mid-gut, is extruded from a small opening above the capitulum during oviposition. It secretes waxy waterproofing substances over the eggs during oviposition, enabling them to withstand desiccation, immersion in water and other adverse environmental conditions.

16.3   LIFE CYCLE

A blood-meal is essential for egg production, and in argasids feeding is mainly nocturnal. Most *female ticks* ingest large blood-meals; soft ticks may increase their weight 12-fold after feeding, while hard ticks take up even more blood (p. 225). After each blood-meal female argasid ticks lay *several* (often 4–6) small egg batches, each comprising 15–100 spherical eggs. Occasionally egg batches may have as many as 300–500 eggs. Because adult ticks can live for many years a female may lay thousands of eggs during her lifetime. Eggs are deposited in or near the resting places of adult ticks, such as in cracks and crevices in the walls, floors and furniture of houses, or in mud, dust and debris, in rodent holes or in the more exposed resting or sleeping places of wild animals and birds.

Eggs usually hatch after 1–3 weeks, but because they have been coated during oviposition with a protective waxy secretion from Gene's organ (see above) they can remain viable for many months under adverse climatic conditions.

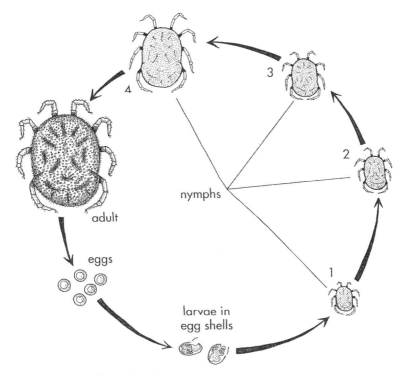

Figure 16.4 Life cycle of *Ornithodoros moubata* showing larvae retained in eggshells and four nymphal stages.

Both argasid and ixodid ticks have a *hemimetabolous* life cycle, that is eggs hatch to produce six-legged *larvae* which superficially resemble the adults, and which moult to produce eight-legged *nymphs* which resemble the adults even more closely. In argasid ticks the six-legged larva (Fig. 16.3) is usually very active and searches for a host from which to take a blood-meal. The capitulum projects from the body and is visible from above. Blood-feeding may last 20–30 minutes, but in some species as much as several days, after which the engorged larva drops to the ground and after a few days moults to produce an eight-legged nymph. The nymph seeks out a host and feeds for about 20–35 minutes before it falls to the ground. Argasid ticks usually have *four or five* nymphal instars (Fig. 16.4), but up to seven in some species. Each nymphal stage requires a blood-meal before it can proceed to the next nymphal instar. *Adults* usually feed on a host for 20–35 minutes.

Larvae of *Ornithodoros moubata* differ from most other argasid ticks because they do not take blood-meals but remain within their eggshells after hatching, moulting to produce first-instar nymphs which crawl from the eggshells to seek blood-meals.

The duration of the *life cycle*, from egg hatching to adult, depends on the species of tick, temperature and the availability of blood-meals, but in argasids is typically about 6–12 months. Adult ticks can *live* for many years, up to 12–20 years in the laboratory; the record for a tick is 25 years! In the absence of suitable hosts adults can survive up to 12 years without a blood-meal. In argasid ticks mating usually occurs away from the host, such as on the ground or amongst vegetation.

The distribution of the larvae, nymphs and adults of argasid ticks is usually patchy and restricted to homes of their hosts. Such ticks are called nidicolous as apposed to the non-nidicolous ixodid ticks. Species which commonly feed on people, such as *O. moubata* in Africa, are found around human settlements, especially in village huts. They can also be found in livestock shelters, chicken sheds, animal burrows and caves, especially in dry areas. However, in many parts of Africa argasid ticks appear to be becoming uncommon. This may be due to changes in lifestyle, in particular to the increased numbers of people sleeping on beds raised from the floor, which makes it more difficult for the ticks to feed on humans.

Because larvae and all nymphal instars and adults take blood, but nevertheless remain attached for only relatively short periods, many hosts, comprising different individuals and often different species, are fed upon during their life cycle. Argasid ticks are consequently referred to as '**many-host**' or 'multi-host' ticks.

16.4   MEDICAL IMPORTANCE

16.4.1 Tick-borne relapsing fever

Soft ticks can inflict painful bites and although they can sometimes cause tick paralysis this condition is more common following prolonged feeding by hard ticks (p. 228).

Tick-borne relapsing fever is the only important disease transmitted to humans by soft ticks. The infection occurs throughout most of the tropics, subtropics and in many areas of the temperate region such as North America and Europe, but is absent from Australia and New Zealand.

There are about 14 species of *Borrelia*, each having different geographical distributions, that cause *Ornithodoros*-transmitted relapsing fevers. The most common is *B. duttonii*, found in sub-Saharan Africa and transmitted by *O. moubata*. In other geographical areas different ticks in the *O. moubata* complex transmit different species of *Borrelia*.

Spirochaetes ingested with a blood-meal multiply in the mid-gut, penetrate its wall and pass across into the haemocoel where they can be found after 24 hours. In the *haemocoel*, the spirochaetes multiply enormously and invade nearly all tissues and organs of the tick's body. After 3 days they begin to arrive in the *salivary glands*, the *coxal organs* and *ovaries*. In *O. moubata* nymphs the salivary glands are more heavily infected than

Table 16.1. *Summary of principal features distinguishing soft and hard ticks*

Argasid ticks (Soft ticks)	**Ixodid ticks** (Hard ticks)
Morphology	
No scutum (shield)	Scutum (shield) on larvae, nymphs and adults. Females with small, and males with large scutums
Mouthparts (capitulum) not visible dorsally in nymphs and adults, but seen in larvae	Mouthparts (capitulum) visible dorsally in larvae, nymphs and adults
Palps leg-like; chelicerae have smooth sheaths	Palps club-shaped; chelicerae have denticulate sheaths
Coxal organs present	Coxal organs absent
Life cycle	
Eggs laid in small batches (15–100)	Eggs laid in one large batch of many thousands of eggs
4–5 nymphal stages (8-legged)	Only one nymphal stage (8-legged)
Adults blood-feed rapidly, on hosts for only 20–35 minutes, but feed on several separate occasions	Adults feed slowly, on hosts for 1–4 weeks, but females feed only once
Multi-host ticks, usually about 6 hosts	Usually 2- or 3-host ticks
Ticks found mainly in or around homes of host, disperse little	Ticks attach to host for long time hence can disperse considerable distances
Diseases	
Vectors of tick-borne relapsing fever	Vectors of tick-borne typhuses, Lyme disease and many viruses. Cause tick paralysis

those of the adults, in which the infection tends to die out. In contrast the coxal organs of the nymphs are usually only lightly infected whereas those of the adults become heavily infected. When either nymphs or adults of *O. moubata* feed on humans, or some other host, saliva is injected intermittently into the bite and spirochaetes can be introduced by this route, especially by nymphal ticks. During feeding excess body fluids are filtered from the haemocoel by the coxal organs and in infected ticks, especially adults, the coxal fluids contain spirochaetes ingested with a previous blood-meal. These can enter the host through the puncture of the tick's bite or through intact skin. Humans can therefore become infected with *B. duttonii* by either the bite of *O. moubata* or the coxal fluids, or both.

In other *Ornithodoros* species the coxal organs tend to excrete excess fluids only when the ticks have left their host, and consequently transmission by these species is mainly by tick bite. In no species of *Ornithodoros* is infection spread by faeces.

The *tick* is usually regarded as the most important reservoir, especially as there is *transovarial* transmission. That is the ovaries of adult female ticks become infected with spirochaetes which are then passed on to the eggs so that the newly hatched larvae (but not of *O. moubata* because they do not blood-feed), and all nymphal instars and adults of both sexes, are infected. Thus, although nymphs and adults may not have fed on a host infected with *B. duttonii* they can nevertheless transmit the infection to other hosts. Transovarial transmission can be continued for some three to four generations.

Another rather similar and associated method of transmission is *transstadial* transmission. This involves one of the immature stages of a tick becoming infected by biting a host and then passing the infection on to one or more later stages. For example, a larva might become infected by feeding on an infected host and pass the spirochaetes to the nymphs and adults, or the infection might start with a nymph and be passed to subsequent nymphal instars and the adults. In all cases transovarial transmission can follow.

Although *B. duttonii*-transmitted relapsing fever is not a zoonosis, other tick-borne relapsing fevers, such as *Borrelia hermsii* transmitted by the tick *O. hermsi* in the Americas, are zoonoses, usually with rodents as reservoir hosts.

16.4.2 Q fever
Q fever is transmitted mainly by ixodid ticks although argasid ticks can also be vectors. See page 233 for an account.

16.4.3 Viruses
More than 100 arboviruses are transmitted by ticks, but only about 30 have been isolated from soft ticks, and very few infect people. Soft ticks, in marked contrast to hard ticks, are not important medical vectors of arboviruses to humans.

16.5   CONTROL
Methods used for removing ticks from their hosts are described in Chapter 17 (p. 234).

Suitable repellents that can be applied to the skin include DEET, dibutyl phthalate and dimethyl carbamate. Alternatively, clothing can be impregnated with these chemicals, or with permethrin or cyfluthrin insecticides (when applied to ticks or mites insecticides are often called acaricides), to help prevent tick infestations.

Houses infested with argasid ticks, such as *Ornithodoros* species, can be sprayed with insecticides such as 3% malathion, 0.5% chlorpyrifos, 0.5% diazinon, 1% pirimiphos methyl, 3% bendiocarb, 1% propoxur or 0.3% permethrin. Care should be taken to spray floors and cracks and crevices in walls and furniture, and other sites where ticks may be resting. In better-built houses where walls are uniformly plastered this usually reduces the numbers of ticks resting in them. When houses have been sprayed with residual insecticides for malaria control there is often a reduction in the numbers of *Ornithodoros*.

FURTHER READING

Evans, G. O. (1992) *Principles of Acarology*. Wallingford: CAB International.

Krantz, G. W. (1978) *A Manual of Acarology*. 2nd edn. Corvallis, OR: Oregon State University.

McCall, P. J. (2001) Tick-borne relapsing fever. In: *The Encyclopedia of Arthropod-Transmitted Infections of Man and Domesticated Animals*, ed. M. W. Service. Wallingford: CABI. pp. 513–6.

McDaniel, B. (1979) *How to Know the Mites and Ticks*. Dubuque, IA: W. C. Brown.

Obenchain, F. D. and Galun, R. (1982) *Physiology of Ticks*. Oxford: Pergamon Press.

Oliver, J. H. (1989) Biology and systematics of ticks (Acari: Ixodoidea). *Annual Review of Ecology and Systematics*, **20**, 397–430.

Sauer, J. R. and Hair, J. A. (1986) *Morphology, Physiology, and Behavioral Ecology of Ticks*. Chichester: Ellis Horwood; New York: Wiley.

Schuster, R. and Murphy, P. W. (eds.) (1991) *The Acari: Reproduction, Development and Life History Strategies*. London: Chapman and Hall.

Sonenshine, D. E. (1991) *Biology of Ticks, volume 1*. New York and Oxford: Oxford University Press.

(1993) *Biology of Ticks, volume 2*. New York and Oxford: Oxford University Press.

See also references to hard ticks at the end of Chapter 17.

17

Hard ticks (Ixodidae)

eye

eye

scutum

festoons

scutum

festoons

MALE   FEMALE

Figure 17.1 Adults of hard ticks: male *Amblyomma* (courtesy of the Natural History Museum, London) and female *Dermacentor*, showing sexual differences. A male has a large scutum while a female ixodid has a small scutum. Note the presence of festoons.

Hard ticks (Ixodidae) have a worldwide distribution, but they occur more frequently in temperate regions than soft ticks (Argasidae). There are about 640 species of hard ticks belonging to 11 genera. (See p. 214 for disagreements on tick classification). Medically the more important genera are *Ixodes*, *Dermacentor*, *Amblyomma*, *Haemaphysalis*, *Rhipicephalus* and *Hyalomma*. Some hard ticks are vectors of typhuses such as Rocky Mountain spotted fever (*Rickettsia rickettsii*) and Meditteranean spotted fever (*R. conorii*), and some can spread Q fever (*Coxiella burnetii*). Many arboviruses, including tick-borne encephalitis, Omsk haemorrhagic fever, Kyasanur Forest disease, Crimean–Congo haemorrhagic fever and Colorado tick fever, are transmitted by hard ticks. They also transmit tularaemia (*Francisella tularensis*), and cause tick paralysis.

17.1   EXTERNAL MORPHOLOGY

Adult hard ticks are *flattened* dorsoventrally and are oval in shape, measuring 2–23 mm in length, size depending on species and whether they are unfed or fully engorged with blood. Females are usually bigger than males, and because they take larger blood-meals they enlarge much more than males during feeding.

The *capitulum* or 'false head' projects forwards from the body and is visible from above (Fig. 17.1), thus distinguishing adult hard (ixodid) ticks from soft (argasid) ticks (Fig. 16.1). Differences in the capitulum also serve to separate the two families. For example, in hard ticks the *palps* are swollen and club-shaped (Fig. 17.2), rather than leg-shaped as in the soft ticks. Unlike

Figure 17.2 Capitulum of an adult ixodid tick showing club-shaped palps with minute fourth segment, and denticulate cheliceral sheaths.

soft ticks the cheliceral sheaths are covered with very small denticles, giving them what is sometimes termed a 'shagreened' appearance. As in argasid ticks both the *hypostome* and *chelicerae* penetrate the host during feeding. The basal part of the capitulum is termed the basis capituli. In hard, but not soft, ticks a cement-like substance produced by the salivary gland 'glues' the mouthparts firmly into the host, and ensures continued attachment during their long feeding times (days to weeks).

The posterior margin of the body in species of the genera *Dermacentor*, *Rhipicephalus* and *Haemaphysalis* has a number of rectangular indentations called *festoons*. However, in fully engorged females these indentations may be difficult to see due to the body's distension with blood.

All hard ticks have a dorsal plate called a dorsal shield or *scutum*, which is absent in soft ticks. In *males* the scutum is large and covers almost the entire dorsal surface of the body (Fig. 17.1), whereas in *females* it is much smaller and is restricted to the anterior part of the body just behind the capitulum (Fig. 17.1). In fully fed females the scutum may be difficult to see because it appears small in relation to the enlarged body and becomes pushed forwards so that it is almost vertical in position. In both sexes of *Dermacentor*, *Amblyomma* and some *Rhipicephalus* species the scutum has so-called enamelled coloured areas, and such ticks are described as *ornate* species. The scutum provides a method of immediately recognizing hard ticks and also of differentiating between the sexes, although medically this may not be very important because *both* sexes take blood-meals and are therefore potential disease vectors. In the larval and nymphal stages the scutum is small in both sexes.

There are four pairs of legs, with each leg terminating in a pair of claws. There are no coxal organs in ixodid ticks. The internal organs are basically as encountered in argasid ticks (p. 215).

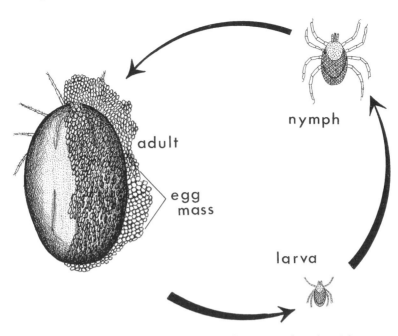

Figure 17.3 Life cycle of an ixodid tick showing a female with a very large egg mass and a single nymphal stage.

17.2 LIFE CYCLE

Both the Ixodidae and Argasidae have *hemimetabolous* life cycles, that is there is incomplete metamorphosis involving a *larval* and *nymphal* stage. There are, however, important differences between the life cycles and ecology of ticks in these two families. *Adult* ixodid ticks remain attached to their hosts for long periods as blood-feeding often lasts 1–4 weeks. When feeding has finished the enormously engorged tick drops from the host to the ground and shelters under leaves, stones, detritus, amongst surface roots of grasses and shrubs, or buries itself in the surface soil. These are termed non-nidicolous ticks in contrast to nidicolous argasids which live in hosts' burrows and homes. The time taken for females to digest their blood-meal and commence laying eggs varies according to species and environmental conditions, especially temperature. Sometimes oviposition begins 3–6 days after the female drops from the host, but egg laying may not begin until several weeks or occasionally months after the end of feeding. *Thousands* (1000–10 000) of small spherical eggs are laid in a gelatinous mass which is formed in front and on top of the tick's scutum (Fig. 17.3). A few species lay as many as 20 000 eggs and the egg mass may become larger than the ovipositing female. Oviposition may last 10 days, or extend over a month or more. As in argasid ticks the eggs are coated with a waxy secretion produced by *Gene's organ*, which in the case of ixodid ticks also helps transfer eggs from the genital opening to the scutum. The ixodid female lays *only* one batch of eggs, after which she dies.

After 10–20 days to a few months six-legged larvae hatch from the eggs. The larvae are minute (0.5–1.5 mm long), and are sometimes called *seed ticks*. On cursory examination they superficially resemble larval mites, but the presence of a *toothed* hypostome immediately identifies them as ticks. After emergence larvae remain inactive for a few days, after which they swarm over the ground and climb up vegetation and cluster at the tips of grasses and leaves. When suitable hosts pass through a tick-infested habitat the larvae respond to stimuli such as carbon dioxide, host odours, warmth, shadows, vibrations and movements by waving their front legs in the air. This host-seeking behaviour exhibited by the larvae, and also by the nymphs and adults, is called *'questing'*. Larvae climb onto their hosts and crawl to their favoured sites for attachment. The chelicerae and hypostome are inserted deep into the skin of the host and the larvae commence blood-feeding. *Larvae* remain attached to their hosts for about 3–7 days before dropping to the ground and sheltering amongst vegetation or under stones. They normally take a few days to digest their blood-meals, but in cooler weather digestion may extend over several weeks. After blood has been digested the larvae remain inactive for a further few days before they moult to become nymphs.

Newly formed eight-legged *nymphs* climb vegetation and behave similarly to the larvae (that is *questing*) in seeking a suitable passing host. *Blood-feeding* lasts 5–10 days after which fully engorged nymphs drop to the ground and shelter under stones or amongst vegetation. They remain quiescent for a few weeks while the blood-meal is digested, after which the nymphs moult to produce male or female ixodid ticks. There is only *one* nymphal stage in the life cycle of ixodid ticks (Fig. 17.3), whereas argasid ticks have several nymphal stages.

Newly formed adults remain more or less inactive for about a week, after which they climb vegetation and start questing for passing hosts. *Adult* female ticks can take enormous blood-meals, ingesting 1–8 ml of blood. Although much of the fluid of this is excreted the weight of an unfed female can increase nevertheless some 200 times after blood-feeding! In contrast male ticks ingest very little blood. Adults shelter under stones and surface vegetation, and amongst roots of plants. Except for *Ixodes* species, mating in ixodid ticks occurs on the host and males may remain on a host for several weeks or months, mating with several females.

17.3   BEHAVIOUR AND HABITS

Many species of ixodid ticks are more or less host-specific. For example some species feed almost exclusively on birds, others on reptiles, or on certain types of mammals such as bats, canids or bovids, but other species feed on almost any available hosts, including people. Diversity of host species may increase the likelihood of disease transmission. *Larvae and nymphs* of many, but not all, ticks seem to have a predilection for small animals such as rodents, cats and dogs, and ground-inhabiting birds, whereas *adults* prefer to feed on cattle, deer, horses and a variety of large and wild mammals. *Humans* are parasitized by the younger stages of hard ticks, rarely by adults.

The *life cycle* of ixodid ticks commonly takes 2–4 years, depending on species and availability of hosts, but may be even longer due to lack of hosts. In temperate regions development may also be prolonged or cease temporarily during winter months. In warm countries development and breeding often continue throughout the year; nevertheless there may be seasonal fluctuations in development rates. *Adult* ticks can withstand starvation for 1–2 years, and may live for 7 years, but usually they live for only about 2 years.

Although ticks can tolerate considerable variations in temperature and humidity, most species are absent from both very dry and very wet areas, but certain *Hyalomma* species occur in arid areas. Microclimatic conditions at soil level are greatly influenced by the amount and type of ground vegetation, and consequently the distribution of different species of ticks is often closely associated with particular types of vegetation.

Both immature and adult ixodid ticks remain on their hosts much longer than do argasid ticks and consequently may be carried many kilometres by their hosts, or even across continents by migrating birds. They are therefore not restricted to their hosts' homes or resting places as are most argasid ticks, but are usually more widely dispersed.

17.3.1 Three-host ticks

The life cycle previously described refers to a three-host tick: that is a different individual host, which may be the same or different species, is parasitized by the larva, nymph and adult, and moulting occurs on the ground. About 95% of ixodid ticks have this type of life cycle, and medically important species of three-host ticks are found in the genera *Ixodes*, *Dermacentor*, *Rhipicephalus*, *Haemaphysalis* and *Amblyomma*. Ticks which feed on three hosts are more likely to become infected with pathogens and be potential vectors of disease than species that feed on one or two hosts.

17.3.2 Two-host ticks

A few species, in genera such as *Hyalomma* and *Rhipicephalus*, are two-host ticks. After the larva has completed feeding it remains on the host and moults to produce a nymph which then feeds on the same host. The engorged nymph drops off, moults and the resultant adult feeds on a different host.

17.3.3 One-host ticks

In a few ticks, such as *Boophilus* species, the larva, nymph and adult all feed on the same host and moulting also takes places on that host. The only stage that leaves the host is the blood-engorged female tick which drops to the ground to lay eggs. One-host ticks are less likely to acquire infections with pathogens than ticks which feed on several hosts, and clearly the only method by which infection can be spread from one host to another by these ticks is by transovarial transmission. One-host ticks are of little or no medical importance, but certain species of *Boophilus* are important vectors of animal infections.

See Table 16.1 (p. 219) for a summary of the differences between hard and soft ticks.

17.4   MEDICAL IMPORTANCE

Table 17.1 very briefly summarizes some of the infections transmitted to humans by hard ticks. More details are found below under the various headings.

17.4.1 Tick paralysis

Female hard ticks, mainly certain species of *Dermacentor* and *Ixodes*, cause a condition known as tick paralysis. Human cases have been reported from North and South America, Europe, Asia, Australia and South Africa. The condition also affects pets and domesticated animals. Symptoms appear 4–7 days after a tick, usually a female, has commenced feeding. There is an acute ascending paralysis affecting firstly the legs with the result that the person cannot walk or stand, and later the arms cannot be moved and there follows difficulty in speaking, swallowing and breathing. Symptoms are painless and there is very rarely any rise in the patient's temperature. Tick paralysis can be confused with paralysis due to poliomyelitis and certain other paralytic infections. Young children, especially up to the age of 2 years, are most severely affected. Death in animals, and in rare cases also humans, can result from respiratory failure. Removal of ticks often results in complete recovery after 48 hours, but in severe cases recovery may take a few days, or even up to about 6 weeks.

Tick paralysis is not caused by any pathogens but by toxins contained in the female tick's saliva which is continually pumped into the host during the tick's long feeding period. Different species of ticks and also different populations of the same species may vary markedly in their ability to produce tick paralysis in humans and animals.

17.4.2 Arboviruses

More than 100 arboviruses have been recovered from ticks, but the important tick-borne viral diseases of humans are spread by hard ticks. All

Table 17.1. *Some infections transmitted to humans by hard ticks*

Disease	Infective agent	Principal tick vectors	Hosts in addition to humans
Tick-borne encephalitis	*Flavivirus*	*Ixodes ricinus* *I. persulcatus*	Rodents[a], insectivores[a]
Omsk haemorrhagic fever	*Flavivirus*	*Dermacentor reticulatus*	Muskrats[a], rodents
Kyasanur Forest disease	*Flavivirus*	*Haemaphysalis spinigera* *H. turturis*	Monkeys[a], deer, cattle
Crimean–Congo haemorrhagic fever	*Nairovirus*	*Hyalomma marginatum* complex	Hares[a], cattle[a], hedgehogs[a], birds, sheep
Colorado tick fever	*Coltivirus*	*Dermacentor andersoni*	Rodents[a]
Rocky Mountain spotted fever	*Rickettsia rickettsii*	*Dermacentor, Amblyomma* and *Rhipicephalus* species	Rodents[a]
Mediterranean spotted fever	*Rickettsia conorii*	*Rhipicephalus sanguineus*	Rodents[a], dogs[a]
African tick-bite fever	*Rickettsia africae*	*Amblyomma* species	Cattle
Q fever	*Coxiella burnetii*	Many ixodid species	Rodents, cattle
Human monocytic ehrlichiosis	*Ehrlichia chaffeensis*	*Amblyomma americanum*	Deer[a]
Lyme disease	*Borrelia burgdorferi*	*Ixodes ricinus* *I. scapularis* *I. pacificus*	Rodents[a], birds[b]
Tularaemia	*Francisella tularensis*	Many ixodid species	Rabbits[a], hares[a], deer[a], beavers[a]
Tick paralysis	Tick toxins	Mainly *Ixodes* and *Dermacentor* species	Cattle, sheep, dogs etc

[a] Main vertebrate reservoir hosts
[b] Suspected reservoir hosts

arboviruses are transmitted by the tick's bite, and *transovarial* transmission usually occurs.

Tick-borne encephalitis (TBE) (*Flavivirus*)

First described in 1932 as Russian spring–summer encephalitis (RSSE), and then discovered after World War II (1939–45) in central Europe and called central European encephalitis (CEE). These two infections are now considered to be the same and are called tick-borne encephalitis (TBE). It is widespread in Europe (except the UK, Benelux countries and Iberian Peninsula), Siberia, northern Asia, China and Japan.

In Siberia eastwards the main vector is *Ixodes persulcatus* while in Europe it is *Ixodes ricinus*. After multiplication in the tick, virus accumulates in the salivary glands, and infection is through the tick's bite. Various small rodents and insectivores (e.g. hedgehogs) in addition to ticks serve as *reservoir hosts*. There is *transstadial* and *transovarial* transmission. Humans are not part of the natural transmission cycle but become accidentally infected by ticks.

TBE virus accumulates in the mammary glands of goats, sheep and cows, and people may become infected by drinking infected unpasteurized milk or eating cheese.

Omsk haemorrhagic fever (OHF) (*Flavivirus*)

The virus causing OHF is antigenically very similar to viruses causing TBE and KFD, and clinical symptoms are rather similar to those caused by these other viruses. OHF occurs in Siberia, such as in the Omsk region. The primary vector is *Dermacentor reticulatus* (formerly called *D. pictus*) which feeds on rodents, especially *muskrats*, which are *amplifying hosts*. Infections acquired from these animals are transmitted *transstadially* to nymphs or adults and then to humans. *Transovarial* transmission also occurs. Muskrat hunters are particularly liable to come into contact with infected ticks, and the disease can also be passed directly to them by the animals' urine and faeces.

Kyasanur Forest disease (KFD) (*Flavivirus*)

KFD was first recognized in 1957 when monkeys were dying in Kyasanur Forest in Karnataka State of southern India and people were also becoming ill and dying. The disease is now found in about 5000 km^2 in and around Kyasanur Forest and is associated with encroachment of people into forests, deforestation for food crops and cattle grazing at the forest edge, all situations having tick foci. The main vectors are species of *Haemaphysalis*, especially *H. spinigera* which transmits the virus to humans, while *H. turturis* maintains animal transmission.

Larval ticks feed on birds and small forest rodents, whereas the nymphal stages feed mainly on monkeys and humans. *Monkeys* are regarded as the principal *amplifying hosts*. Larger mammals such as deer, bison and cattle brought to the forest edge to graze serve as hosts for adult ticks and help maintain large tick populations, but are not reservoirs of viral infection. There is *transstadial*, but apparently no transovarial, transmission in the tick vectors. The epidemiology of KFD is particularly interesting because it shows how changes in people's behaviour, deforestation and agricultural development can lead to changing ecology and disease outbreaks in the human population.

Crimean–Congo haemorrhagic fever (CCHF) (*Nairovirus*)

CCHF virus is recorded from central and southern Europe, the Middle East, Russia, Pakistan, India, China, Madagascar and in Africa from Mauritania to Ethiopia to South Africa. After dengue viruses CCHF virus is the most widely distributed arbovirus, with human infections known from 25 countries and virus isolations from ticks in another 10 countries. The disease is typically enzootic in savannah, steppe and semidesert areas. Transmission is mainly by *Hyalomma* species, such as *H. marginatum marginatum*, but in Africa *H. marginatum rufipes* is the vector. Larval and nymphal ticks feed on birds and small mammals, while adults feed on larger mammals including humans. Hares, hedgehogs and cattle are probably *amplyfying hosts*. Although *birds* are not reservoirs of infection they can spread infected ticks around the world; for example some 5 billion birds fly annually from Europe to Africa, and about half return. Adult ticks attack humans and large mammals such as cattle and sheep.

Transmission is by tick bite and possibly by crushing infected ticks, or by accidental contamination from infected blood during sheep shearing. *Ticks* are regarded as important reservoirs of infection, especially as they can survive starvation for at least 800 days! There is *venereal* transmission, i.e. virus is transmitted from infected male ticks to uninfected females during mating, and then *transovarial* transmission from infected females to their progeny.

Colorado tick fever (CTF) (*Coltivirus*)

CTF occurs in the Rocky Mountain states and South Dakota in the USA and in western Canada. The principal vector is *Dermacentor andersoni*. Larvae and nymphs feed on small mammals such as rabbits, ground squirrels, chipmunks and woodrats, which together with ticks are the main *reservoirs* of infection. Adult ticks feed on larger mammals, such as deer, cattle and people. There is both *transstadial* and *transovarial* transmission.

17.4.3 Rickettsiae

Tick-borne typhuses have an almost worldwide distribution and are caused by 13 species of *Rickettsia*. Ticks are usually regarded as the main *reservoirs* of infection although rodents and other mammals may sometimes by reservoir hosts. There is usually *transovarial* transmission and often *transstadial* transmission. The more important tick-borne typhuses are described briefly below.

Rocky Mountain spotted fever (RMSF)

RMSF, also known as Mexican spotted fever, São Paulo spotted fever, American tick-borne typhus and by several other local names, occurs in North, Central and South America. The causative agent is *Rickettsia rickettsii*. The principal vector in western America is *Dermacentor andersoni* and in eastern USA *D. variabilis*. In South America *Amblyomma cajennense* is the main vector, and this species and *Rhipicephalus sanguineus* are the important vectors in Central America. RMSF is a *zoonosis* with ground squirrels, chipmunks and other small rodents acting as *reservoir hosts* and/or *amplifying hosts*, although the tick itself is considered the main reservoir of infection, especially as infections can persist in overwintering ticks. Dogs are not reservoirs or amplifying hosts, but they can transport infected ticks to human habitations where they may become dislodged and attach to people.

Transmission is normally through the bite of any stage in the life cycle of the tick. An infective tick, however, must remain feeding on a host for at least 2 hours before sufficient rickettsiae are injected into the host for the host to become infected. Consequently, early tick removal may prevent transmission. There is both *transstadial* and *transovarial* transmission.

Mediterranean spotted fever

Also known as boutonneuse fever, Marseilles fever, South African tick typhus, Kenyan tick typhus, Indian tick typhus and Crimean tick typhus. The infective agent is *Rickettsia conorii*. It occurs in the Mediterranean littoral region, Israel, Portugal, Sicily, eastern Russia, India and sub-Saharan Africa. The principal vector is *Rhipicephalus sanguineus*, the dog tick. Transmission is by the tick's bite, and both *transstadial* and *transovarial* transmission occur. Ticks, various rodents and, in contrast to RMSF, dogs can be *reservoir hosts*. Infection can also occur if infected ticks are *crushed* and the rickettsiae rubbed into abrasions or the eyes.

African tick-bite fever

Initially confused with typhus caused by *R. conorii*, but in 1992 the causative agent was named *R. africae*. This form of typhus is common throughout most of sub-Saharan Africa, and also occurs in the West Indies. In both regions vectors are *Amblyomma* species.

Miscellaneous tick-borne typhuses

These include Siberian tick typhus (*R. sibirica*), Queensland tick typhus (*R. australis*) and Japanese or Oriental tick typhus (*R. japonica*) as well as unnamed diseases caused by other *Rickettsia* species.

Q fever

Q fever is a rickettsial zoonotic disease caused by *Coxiella burnetii*. It has a worldwide distribution and is primarily an infection of rodents, other small mammals and domestic livestock. It can be transmitted to people by inhalation of aerolized rickettsia, possibly by drinking contaminated milk, and occasionally by bites of argasid ticks, but mainly ixodid ones. Ticks are probably important in maintaining infections in wild animals and in transmission to domesticated ones. *Transovarial* transmission occurs.

17.4.4 Ehrlichiosis

Species of *Ehrlichia* infect dogs and ruminants, and some species are zoonotic and infect humans. One species, *E. chaffeensis*, causes human monocytic ehrlichiosis (HME), while *E. phagocytophila* and *E. ewingii* parasitize granulocytes and cause human granulocytic ehrlichiosis (HGE). Transmission is by bites of hard ticks. *Transstadial* transmission occurs, and possibly also transovarial. Most human cases occur in the USA, where the disease seems to be increasing in prevalence, but ehrlichiosis has also been reported from Europe, Africa and Venezuela.

17.4.5 Spirochaetes

Lyme disease

Lyme disease (also called Lyme borreliosis or erythema migrans) was first recognized in 1975 in the town of Old Lyme, Connecticut, USA. It is caused by the spirochaete *Borrelia burgdorferi*, one of 11 species within the *B. burgdorferi* complex. Here the parasite will be referred to as *B. burgdorferi*, although in some cases of Lyme disease the infectious agent may be another species, such as *B. afzellii*.

Lyme disease occurs in Europe, Asiatic and European parts of the former USSR, China, Japan, the USA, Canada, and possibly in Africa and Australia but reports from these areas need confirmation. In Europe transmission is by the bite of *Ixodes ricinus* and in Eurasia by *I. persulcatus*. In the eastern USA the vector is *Ixodes scapularis* (= *I. dammini*) whereas in western areas *I. pacificus* is the principal vector. There is both **transstadial** and **transovarial** transmission. Lyme disease is the most commonly reported arthropod-transmitted infection in North America and Europe. In the USA 23 763 cases were reported in 2002.

Lyme disease is a zoonosis. More than 100 animal species have been identified as being infected, but rodents and possibly birds are the most important *reservoir hosts*. Deer although supporting large populations of vector ticks are not reservoir hosts.

The ecology of Lyme disease and reasons for its increased prevalence and the extension of its geographical range are both complex and interesting. It seems that one explanation is that in both North America and Europe people are spending more time in rural areas where there can be large infestations of ticks from deer. Also people are increasingly building homes next to recently cleared land adjacent to forests and this increases their chances of being bitten by ticks. Increased awareness and more extensive serological testing of people may also be partly responsible for the increased numbers of reported cases.

17.4.6 Tularaemia

Tularaemia is a bacterial disease caused by *Francisella tularensis*. It occurs throughout the northern hemisphere, being most common in the USA and southern regions of Russia. It infects mainly rabbits and hares, but also other rodents, beavers and deer, all of which can be *reservoir hosts*. The infection is spread by a variety of direct contact methods such as handling infected live animals or carcasses, drinking contaminated water, eating raw and uncooked meats and also by the bites of various hard ticks. The tabanid fly *Chrysops discalis* can also be a vector (p. 114).

17.5   CONTROL

Personal protection methods include application of repellents such as DEET, dibutyl phthalate or dimethyl carbamate. Clothing, especially socks and trousers, can be sprayed with repellents or with pyrethroids such as permethrin or cyfluthrin.

Many methods are advocated for removal of ticks from their hosts including coating them with vaseline, medicinal paraffin or nail varnish, all of which block their spiracles thus preventing ticks from breathing. But it may be several hours before such ticks will withdraw their mouthparts and this is usually unacceptable, especially as *rapid removal* of ticks often reduces the chances of disease transmission (e.g. RMSF). With soft (argasid) ticks more rapid removal is sometimes achieved by dabbing them with chloroform or some other anaesthetic. But this method rarely works with hard ticks because they are attached to their hosts with a type of 'saliva cement' which prevents rapid withdrawl of their mouthparts. The best method is to grasp the *capitulum* ('false head') with blunt forceps and slowly pull the tick out, and then to apply an antiseptic to the bite wound.

Many species of hard ticks transmit livestock infections and world-wide regular dipping of sheep and cattle in acaricidal baths, or spraying them with insecticides (acaricides) is practised. Such intensive use of insecticides has resulted in ticks becoming resistant to some organochlorine, organophosphate, carbamate and pyrethroid insecticides. Despite this, insecticides are still needed to control medically important ticks. For example, solutions of 5% malathion, 0.1% dioxathion, 0.2% naled, 1% coumaphos or the carbamate 1% carbaryl can be applied to dogs to kill their ticks, which may otherwise attach to humans. Alternatively, dusts of 0.5% coumaphos, 3–5% malathion, 1% trichlorphon or 5% carbaryl can be used. Floors of houses, porches, verandahs and other sites where infected pets sleep should be sprayed with insecticides as outlined on page 221 for control of endophilic soft ticks.

Ticks in gardens, yards and nearby fields can be controlled by ground applications of 1–8 kg/ha of diazinon, chlorpyrifos, carbaryl, propoxur, permethrin or other pyrethroids. Pellet formulations are best because they more easily penetrate vegetative ground cover and reach the microhabitats of the ticks; a single application may be effective for 6–8 weeks. ULV (ultra-low-volume) spraying with these chemicals reduces dosage rates to about 0.34–1.2 litres/ha, but there may be environmental objections to such blanket-type spraying of vegetation.

FURTHER READING

Camicas, J.-L., Hervy, J.-P., Adam, F. and Morel, P.-C. (1998) *The Ticks of the World* (Acarida, Ixodida): *Nomenclature, Described Stages, Hosts, Distribution.* Paris: Editions de l'ORSTOM.

Gothe, R., Kunze, K. and Hoogstraal, H. (1979) The mechanisms of pathogenicity in the tick paralyses. *Journal of Medical Entomology*, **16**, 357–69.

Gray, J S., Kahl. O., Lane, R. S. and Stanek, G. (2002) *Lyme Borreliosis: Biology, Epidemiology and Control*, Wallingford: CABI.

Hoogstraal, H. (1966) Ticks in relation to human diseases caused by viruses. *Annual Review of Entomology*, **11**, 261–308.

(1967) Ticks in relation to human diseases caused by *Rickettsia* species. *Annual Review of Entomology*, **12**, 377–420.

(1978) Tickborne diseases of humans: a history of environmental and epidemiological changes. In: *Medical Entomology Centenary Symposium Proceedings*, ed. S. Willmott. London: Royal Society of Tropical Medicine and Hygiene. pp. 48–55.

(1981) Changing patterns of tickborne diseases in modern society. *Annual Review of Entomology*, **26**, 75–99.

Lane, R. S., Piesman, J. and Burgdorfer, W. (1991) Lyme borreliosis: relation of its causative agent to its vectors and hosts in North America and Europe. *Annual Review of Entomology*, **36**, 587–609.

Needham, G. R. and Teel, P. D. (1991) Off-host physiological ecology of ixodid ticks. *Annual Review of Entomology*, **36**, 313–52.

Parola, P. and Raoult, D. (2001) Tick-borne typhuses. In: *The Encyclopedia of Arthropod-Transmitted Infections of Man and Domesticated Animals*, ed. M. W. Service. Wallingford: CABI. pp. 516–24.

Reed, G. H. (1993) Lyme disease and other tick-borne diseases. *Journal of Environmental Health*, **55**, 6–10.

Sonenshine, D. E., Lane, R. S. and Nicholson, W. L. (2002) Ticks (Ixodida) In: *Medical and Veterinary Entomology*, ed. G. Mullen and L. Durden. Amsterdam: Academic Press. pp. 517–58.

Sonenshine, D. E. and Mather, T. N. (eds.) (1994) *Ecological Dynamics of Tick-Borne Zoonoses*. New York, NY: Oxford University Press.

See also references to soft ticks at the end of Chapter 16.

18

Scabies mites (Sarcoptidae)

Mites are not insects but like ticks belong to the subclass Acari. They can be distinguished from ticks by the *absence* of teeth on the hypostome of the mouthparts and in having *setae* (bristles) on the body and legs. But the principal medically important species (scabies mite, scrub typhus mite, house-dust mite and follicle mite) can most readily be recognized by their characteristic shapes.

Sarcoptes scabiei, the scabies or itch mite, occurs on people worldwide. Morphologically they are inseparable from *S. scabiei* infesting wild and domesticated animals, including dogs, horses and pigs, which are considered to be the same species as those found on people but physiologically adapted for life on non-human hosts. In animals they cause the condition known as mange. Mites from animals very rarely infect humans, but if they do the infection can persist for several weeks.

Scabies mites are not vectors of any disease but cause conditions known in people as scabies, acariasis, crusted or Norwegian scabies. There are at least 300 million scabies cases annually.

18.1   EXTERNAL MORPHOLOGY

The female mite (0.30–0.45 mm) is just visible without the aid of a hand lens. It is whitish and disc-shaped. Dorsally the mite is covered with numerous small *peg-like spines* and a few bristles (setae), and both dorsally and ventrally there are series of lines across the body giving the mite a striated appearance (Fig. 18.1). Adults have four pairs of short and cylindrical legs divided into five ring-like segments. The first two pairs of legs end in short stalks called *pedicels* which terminate in thin-walled roundish structures often termed '*suckers*'. In *females* the posterior two pairs of legs do not have 'suckers' but end in long and very conspicuous bristles. There is no distinct head, but the short and fat palps and pincer-like *chelicerae* of the mouthparts protrude anteriorly from the body.

Adult male scabies mites are only 0.20–0.25 mm long, and apart from their small size may also be distinguished from females by the presence of '*suckers*' on the last pair of legs (Fig. 18.2).

18.2   LIFE CYCLE

Scabies was for a long time associated with poverty, overcrowding and poor hygiene, but people from all socioeconomic backgrounds can be infested with mites.

Female mites dig and eat their way into the surface layers of the skin – the stratum corneum – selecting places where the skin is thin and wrinkled, such as between the fingers and on the wrists, elbows, feet, penis, scrotum, buttocks and axillae. The majority (63%) of mites are found on the hands and wrists and about 11% on the elbows. In women mites may be found burrowing beneath and around the breasts and nipples. Occasionally mites are found on the face and scalp, especially in the postauricular fold, and

Figure 18.1 Dorsal view of an adult female scabies mite (*Sarcoptes scabiei*).

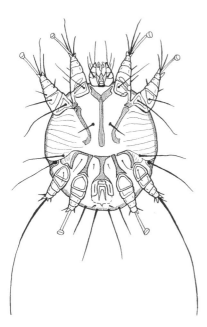

Figure 18.2 Ventral view of the smaller adult male scabies mite showing 'suckers' on the hind-legs (absent on hind-legs of females).

on other parts of the body. The greatest number of mites on children up to a year old are often found on the feet. When females have burrowed into the superficial layers of the skin they excavate winding tunnels at the rate of about 0.5–5 mm per day, which are seen on the skin as very thin twisting lines measuring a few millimetres to 1–2 cm. Mites feed on liquids oozing from dermal cells they have chewed. A female mite lays 1–3 eggs a day in her tunnel.

Eggs hatch after 3–4 days, and small six-legged *larvae* emerge which look like miniature adults. They crawl out of the tunnels onto the surface of the skin where a large number die, but a few succeed in either burrowing into the stratum corneum or entering a hair follicle to produce not a tunnel but a small pocket called a *'moulting pocket'*. After 3–4 days the larva moults in the pocket to produce an eight-legged *nymph* (protonymph) which after 2–3 days moults to produce a second nymph (tritonymph). In mites destined to become females this second nymphal stage (tritonymph) is considerably larger than that of males. After a further 2–3 days the second nymphal stage (tritonymph) moults to become either a male or a female adult. After mating the female adult increases in size and is sometimes referred to as an ovigerous or mature female. Only after fertilization does the female commence to burrow through the skin, and after 3–5 days begins to lay eggs in her tunnel. Female mites rarely leave their tunnels, but contrary to previous beliefs dislodged healthy female mites can sometimes be found on bed linen and on clothing.

Adult male mites are about half the size of ovigerous females and can be found either in very short burrows (usually less than 1 mm) or in small pockets in the skin. However, they probably spend most of their life wandering around on the surface of the skin seeking out female mates.

The *life cycle* from egg to adult usually takes 10–14 days. Female mites may live for 4–6 weeks on humans, but away from their hosts for only 2–5 days.

Scabies is a contagious complaint which is transmitted only by close contact. It is therefore a family disease spreading amongst those living in close association, especially when they share the same bed. Scabies can be spread between courting couples who are habitually holding hands. It appears that the actual *transfer* of mites from person to person takes about 15–20 minutes of close contact. The incidence of scabies often increases during wars and disasters, such as earthquakes, floods and famines, when people are sleeping and living in very overcrowded situations. It is also possible to become infected by sleeping in a bed formerly occupied by an infected person, but this probably rarely happens. As mites may remain alive on discarded clothing for a few days it is possible that some may be able to infest a new host, but such indirect transmission seems very rare. Consequently it may not be worthwhile routinely fumigating or sterilizing clothing or bedclothes to prevent scabies spreading, except in epidemics, or in cases of *crusted scabies* (Norwegian scabies). Ten minutes at 50 °C will kill the mites, as will laundering clothing.

Increases in the incidence of scabies often appear to go in 15–20 year cycles, probably due to fluctuations in levels of immunity in the human population, although this explanation has been disputed.

18.3   RECOGNITION OF SCABIES

Scabies can be diagnosed by detection of the female mite's narrow twisting tunnels; faeces deposited in these tunnels may be visible through the skin as dark pepper-like spots. Unfortunately tunnels are not always easy to see, especially on dark-skinned people. However, they can often be made visible by smearing a drop of ink on suspected areas of skin infestations; after a few minutes the ink seeps into the tunnels and when the excess is wiped off they become visible. Alternatively if liquid tetracycline is used tunnels fluoresce bright yellow-green under ultraviolet light. Surface layers of skin at the end of the tunnels can be gently scratched away with sharp dissecting needles and the mites, which usually readily adhere to the points of the needles, removed and examined under a ×50 magnification. A cruder procedure is to scrape the affected skin areas with a scalpel blade or razor and to examine the scrapings.

The *average* number of mature adult female scabies mites found on a person is about 11; most patients have 1–15 mites, and only about 3% have more than 50 mites.

18.3.1 The scabies rash

The scabies rash is a papular eruption that occurs mainly on areas of the body not infected with burrowing mites, such as the buttocks and around the waist and the shoulders, but the rash can also occur on other parts of the body such as the arms, calves and ankles. It does not appear on the head, centre of the chest or back, nor on the palms of the hands or soles of the feet. The *rash* is in response to the patient being sensitized, that is the rash is an allergic reaction produced by the mites. Frequently patients are unaware they have mites until a rash appears.

When a person is infected for the *first time* with mites the rash does not usually appear until 4–6 weeks, although in exceptional cases the rash may occur within 2 weeks. However, in those who have previously been infected a rash may develop within 2–4 days after reinfection. Severe *pruritus* soon develops and causes vigorous and constant scratching, especially at night and after hot baths. Scratching frequently causes secondary bacterial infections, which may be quite severe resulting in boils, pustules, ecthyma, eczema and impetigo contagiosa. Such complications may hinder the detection of mites and consequently correct diagnosis of scabies may not be made. The seriousness of the symptoms is not always directly related to the number of mites, and severe reactions may be found on people harbouring few mites. The rash may persist for 2–3 weeks after all scabies mites have been destroyed.

A condition known as *crusted* or Norwegian scabies is rare but highly contagious. It is caused by vast numbers of mites, sometimes more than 1000, in exfoliating skin scales, and is characterized by the formation of thick keratogenous crusts over the hands and feet and scaling eruptions on other parts of the body. There is, however, much less pronounced itching. It is unclear why the condition develops, but it seems to be due to a loss of immunity which allows the establishment of enormous numbers of mites. This hypothesis is supported by the finding that the development of crusted scabies has sometimes been associated with the extensive use of corticosteroids, with HIV patients and with those mentally impaired. In such people irritation is reduced and consequently there is less scratching, an act that probably helps in the removal and destruction of mites.

18.4   TREATMENT OF SCABIES

Previously before treatment a patient was given a hot bath and vigorously scrubbed with a brush, but this is not very effective at removing or killing the mites. However, because many, but certainly not all, patients with scabies are dirty, an ordinary bath before treatment may be advisable for general hygienic reasons. But if large numbers of patients need to be treated, such as in epidemic situations, bathing may not be practical.

A 20–25% benzyl benzoate emulsion can be painted on a patient from the neck downwards, and after allowing some 5–10 minutes for this application to dry the patient can re-dress. In most cases a single treatment should give complete eradication of the mites, but a repeat treatment on the third day may be advisable. A mild or severe stinging reaction may result from the use of benzyl benzoate, which is more likely to occur in young children.

Sulphur preparations have been used for years, such as Mitigal which is painted undiluted over the body from the neck downwards. A single treatment should be 100% effective, but a mild form of dermatitis may be produced in some people. Tetmosol is another sulphur compound sometimes used to treat scabies but it acts slowly and usually about three treatments 24 hours apart are recommended for a complete cure. It is therefore of limited use in mass treatments. Crotamiton (Eurax) applied as a 10% cream or lotion is a very safe treatment, but two to five daily applications are needed.

Although the above treatments are effective better treatment for scabies consists of applying more modern acaricides (insecticides). The preferred acaracide is 5% permethrin cream (Elimite, Lyclear), with 1% (0.5% in infants) aqueous malathion as second choice. 1% HCH cream or lotion (Kwell) is also effective, although it has been withdrawn in some countries. Resistance to HCH, permethrin and malathion has been documented. Although single treatments with the above acaricides have a high success rate a *second* application 7 days later, if this is possible, ensures a complete cure.

Ivermectin given as a single oral dose of 200 mg/kg body weight is effective in killing scabies mites, but there are reports of resistance.

Methods aimed at killing mites will do little to immediately alleviate the irritation caused by the rash, and consequently patients continue to complain of itchiness until the irritation eventually disappears. Separate medical treatment may be necessary if secondary infections have become established.

With a highly contagious condition such as scabies it is important to treat all the family or community living in close association, not just the individual diagnosed with mites, otherwise reinfestations will soon occur.

FURTHER READING

Arlian, L. G. (1989) Biology, host relations, and epidemiology of *Sarcoptes scabiei*. *Annual Review of Entomology*, **34**, 139–61.

Cox, N. H. (2000) Permethrin treatment in scabies infestation: importance of the correct formulation. *British Medical Journal*, **300**, 37–8.

Daisley, H., Charles, W. and Suite, M. (1993) Crusted (Norwegian) scabies as a pre-diagnostic indicator for HTLV-1 infection. *Transactions of the Royal Society of Tropical Medicine and Hygiene*, **87**, 295.

Glaziou, P., Cartel, J. L., Alizieu, P., Briot, C., Moula-Pelat, J. P. and Martin, P. M. V. (1993) Comparison of ivermectin and benzyl benzoate for treatment of scabies. *Tropical Medicine and Parasitology*, **44**, 331–3.

Marliere, V., Roul, S., Labreze, C. and Taieb, A. (1999) Crusted (Norwegian) scabies induced by use of topical corticosteroids and treated successfully with ivermectin. *Journal of Pediatrics*, **135**, 122–4.

Meinking, T., L. Burkhart, C. N. and Burkhart, C. G. (1999) Ectoparasitic diseases in dermatology: reassessment of scabies and pediculosis. *Advances in Dermatology*, **15**, 77–108.

Meinking, T. L. and Elgart, G. W. (2000) Scabies therapy for the millennium. *Pediatric Dermatology*, **17**, 154–6.

Mellanby, K. (1943) *Scabies*. Middlesex: E. W. Classey. (Reprinted 1972 with minor additions.)

Mullen, G. and O'Connor, B. M. (2002) Mites (*Acari*). In: *Medical and Veterinary Entomology*, ed. G. Mullen and L. Durden. Amsterdam: Academic Press. pp. 449–516.

Orkin, M. and Maibach, H. T. (eds.) (1985) *Cutaneous Infestations and Insect Bites*. New York, NY: Marcel Dekker. See Chapters 1–18 on scabies.

Paller, A. S. (1993) Scabies in infants and small children. *Seminars in Dermatology*, **12**, 3–8.

Van Neste, D. J. J. (1986) Immunology of scabies. *Parasitology Today*, **2**, 194–6.

19

Scrub typhus mites (Trombiculidae)

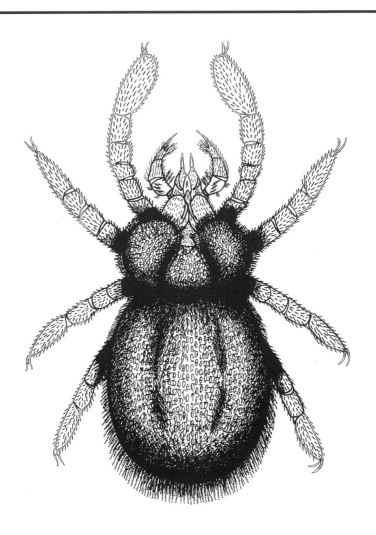

There are some 3000 species of trombiculid mites belonging to several genera but only about 20 species are medically important because they commonly attack people. The family Trombiculidae has a more-or-less worldwide distribution, but the medically important species, such as *Leptotrombidium deliense*, *L. akamushi* and *L. fletcheri*, which are vectors of scrub typhus *Orientia* (= *Ricksettsia*) *tsutsugamushi*, are found in Asia and the Pacific regions.

Other trombiculid mites in many parts of the world cause itching and a form of dermatitis known as scrub itch, autumnal itch or trombidiosis. In northern Europe larvae of *Neotrombicula autumnalis* (**harvest mites**) and in North America and parts of Central and South America larvae of *Eutrombicula alfreddugesi* (**red bugs**) commonly attack people and cause considerable discomfort. Larvae of trombiculid mites are also often called chiggers or scrub typhus mites.

19.1 EXTERNAL MORPHOLOGY

19.1.1 Adults and nymphs

Adults are small (1–2 mm), usually reddish and covered dorsally and ventrally with numerous *feathered hairs* giving them a velvety appearance. There are four pairs of legs ending in paired claws. The body is distinctly constricted between the third and fourth pairs of legs giving it an outline resembling a *figure of eight*. Palps and mouthparts project in front of the body and are clearly visible from above (Fig. 19.1).

The nymph resembles the adult but is smaller (0.5–1.0 mm) and the body is less densely covered with feathered hairs.

Neither adults nor nymphs are of direct medical importance; they do not bite humans or animals but feed on small arthropods and their eggs. Only the larvae are parasitic and hence disease vectors.

19.1.2 Larvae

Larvae are very small (0.15–0.3 mm), but after engorging may increase sixfold in size. They are usually reddish or orange but may be pale yellow or straw-coloured. The three pairs of legs terminate in a pair of relatively large claws. Both legs and body are covered with fine *feathered hairs*. The five-segmented *palps* and *mouthparts* are large and conspicuous, giving the larvae the appearance of having a false head (Fig. 19.2). Dorsally on the anterior part of the body there is a rectangular or pentagonal-shaped **scutum** bearing three to six setae. However, as the scutum is weakly sclerotized it is often difficult to see under the microscope, unless the light is correctly aligned. More easily detected is a pair of eyes on either side of the scutum. In medically important species there are *five feathered setae* on the scutum and in addition a pair of specialized feathered hairs known as *sensillae* which arise from distinct bases. The combination of a body covered

Figure 19.1 Dorsal view of an adult scrub typhus mite (*Leptotrombid-ium* species).

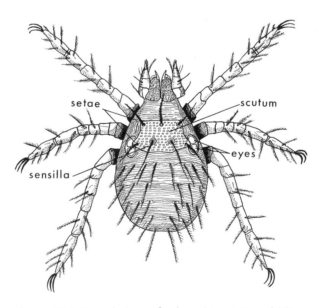

Figure 19.2 Dorsal view of a larval *Leptotrombidium* mite. Note the scutum has a pair of sensillae and five scutal setae.

with feathered hairs, five scutal hairs, a pair of flagelliform sensillae and large pigmented eyes distinguishes larvae of *Leptotrombidium* from larvae in other mite genera.

19.2   LIFE CYCLE

The life cycle of trombiculid mites is complex and unfortunately several different names have been used for the various larval and nymphal stages (see summarized life cycle on p. 248).

Adult trombiculid mites are not parasitic but live in the soil, feeding on a variety of small soil-inhabiting arthropods and their eggs. Females lay 1–5 spherical eggs each day on leaf-litter and the surface of damp but well-drained soil, such as river banks, scrub-jungle, grassy fields and neglected gardens. In hot climates egg laying continues uninterrupted for a year or more, but in cooler areas of South-east Asia, including Japan, oviposition ceases during the cooler months and adults enter into partial or complete hibernation.

After about 4–7 days the eggshell splits, but the six-legged larva does not yet emerge but remains within the eggshell and is called the ***deutovum***. After about 5–7 days the larva crawls out of the eggshell and usually becomes very active, swarms over the ground and climbs up grasses and other low-lying vegetation. Larvae attach themselves to birds and mammals (especially rodents) and also people, who are walking through infested vegetation. When on a suitable host larvae congregate where the skin is soft and moist, such as the ears, genitalia and around the anus. On people larvae seek out areas where clothing is tight against the skin, such as around the waist or ankles.

Larvae piece the host's skin with their powerful mouthparts and inject saliva into the wound which causes disintegration of the cells. Larvae do not suck up blood, but feed on lymph and other fluid and semidigested tissues. The repeated injection of saliva into the wound produces a skin reaction in the host and the formation of a peculiar tube-like structure extending vertically downwards in the host's skin, which is known as the ***stylostome*** or ***histiosiphon***. Some trombiculid mites remain attached to their hosts for as long as 1 month, but the *Leptotrombidium* vectors of scrub typhus remain on people for only about 3–5 days. The engorged larva drops to the ground and buries itself just below the surface of the soil or underneath debris (Fig. 19.3).

Having concealed itself the larva becomes inactive and this stage is known as the ***protonymph***. After 7–10 days the protonymph moults to produce an eight-legged reddish ***deutonymph*** covered with feathered hairs. These nymphs are not parasitic, but feed on soil-inhabiting arthropods. After a few days to about 2 weeks the deutonymph ceases feeding and becomes inactive and is called a ***tritonymph***, which after about another 14 days moults to give rise to an adult. Adults resemble the nymphs but

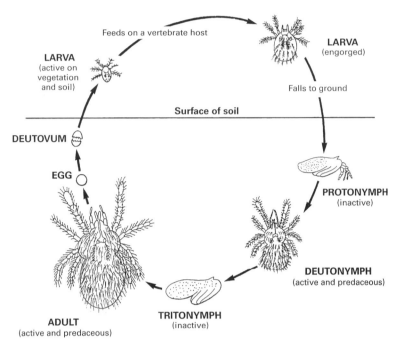

Figure 19.3 Life cycle of a *Leptotrombidium* mite (courtesy of the Natural History Museum, London).

are larger, and like them are free-living and feed on small soil-inhabiting animals.

The life cycle (Fig. 19.3) usually takes 40–75 days but may be as long as 8–10 months. The stages in the life cycle can be summarized as follows (the inactive stages are bracketed, and alternative names are in italics):

Egg → (Deutovum, *prelarva*) → Larva → (Protonymph, *first nymphal stage, nymphochrysalis*) → Deutonymph, *second nymphal stage* → (Tritonymph, *third nymphal stage, preadult, imagochrysalis*) → Adult

19.3   ECOLOGY

Free-living nymphs and adults of *Leptotrombidium* have specialized ecological requirements. For example, habitats must contain sufficient numbers of suitable arthropods to serve as food for nymphs and adults, and must also be traversed by small mammals so that larvae will have hosts on which to feed. **Rodents** such as species of *Rattus*, *Apodemus* and *Microtus* and insectivores such as shrews (*Suncus* species) and tree shrews (*Tupaia* species) are very important larval hosts. Domestic rats play little or no part in the ecology or epidemiology of scrub typhus. In addition to hosts such as these, which maintain mite populations in an area, other more-or-less incidental hosts, including birds, may be important in aiding the dispersal of larvae to other areas.

Typically areas supporting large populations of mites arise from people clearing away vegetation for agricultural purposes or collection of firewood, resulting in the creation of *scrub* vegetation, from which both vector and infection derive their names. Such habitats favouring the survival and development of *Leptotrombidium* mites are called '*mite islands*'. These can be very small, often just a few square metres, or they can comprise several square kilometres. This can result in very patchy and isolated distributions of *Leptotrombidium* mites and areas where people are exposed to infection.

19.4   MEDICAL IMPORTANCE

19.4.1 Nuisance

Several species of Trombiculidae attack humans in temperate and tropical regions of the world. In northern Europe the main pest is the harvest mite (*Neotrombicula autumnalis*), while in the USA it is the red bug (*Eutrombicula alfreddugesi*). Although these mites do not transmit infections they can nevertheless cause intense itching and irritation, commonly referred to as 'harvest-bug itch', 'autumnal itch' or 'scrub itch'. Larval mites commonly attack the legs. If they are forcibly removed, their mouthparts frequently remain embedded in the skin and this may promote further irritation. People usually become infested with these mites after walking through long grass or scrub vegetation, especially in the autumn or summer.

19.4.2 Scrub typhus

The causative organism of scrub typhus is *Orientia tsutsugamushi* and the disease is commonly known as scrub typhus, mite-borne typhus, Japanese river fever, chigger-borne rickettsiosis or tsutsugamushi disease. The disease is restricted to the Asia–Pacific area, extending from the Primorye region of Siberia, through Pakistan and India to Myanmar, Indonesia, Malaysia, Thailand, South-east Asia, China, Taiwan, Philippines, Japan, Papua New Guinea, New Zealand, north-east Australia and neighbouring south-west Pacific islands. Although mostly reported from low-lying areas, infections occur up to 1000 m in many areas, and have been reported up to about 2000 m in Taiwan and 3200 m in the Himalayas.

During World War II (1939–45) the incidence of scrub typhus in troops in the Asia–Pacific area was second only to malaria. More than 40 species of mites are known or suspected as being vectors, but only about seven are important vectors. Over most of the distribution of the disease the principal vector is *L. deliense*, but *L. akamushi* is the chief vector in Japan and *L. fletcheri* is the main vector in Malaysia.

People become infected following the bite of infected larval trombiculid mites encountered when visiting or working in areas having so-called mite islands, that is patches of vegetation harbouring large numbers of host-seeking larvae. The disease is often associated with 'fringe habitats', that is habitats separating two major vegetation zones such as forests and plantations, because these areas are often heavily populated with rodent hosts. The risk of scrub typhus transmission is often related to the number of areas of different types of vegetation, that is habitat diversity.

Because larval mites attach themselves to only a single host during their life cycle the infection cannot be spread by larvae feeding on one infected host (e.g. humans) and then another. Infection acquired by mites feeding on hosts with rickettsiae is passed on to the free-living nymphal stages and then to the free-living adults. When the female lays her eggs they are infected with rickettsiae and this infection is passed on to the emerging larvae. So, although the larvae have not previously fed on humans they are already infected and consequently transmit the disease to their hosts (humans or rodents) when they feed for the first and only time. In addition to such *transovarial* transmission there is also *co-feeding* transmission. That is, when a larva is feeding very near an infected larva which is feeding, the rickettsiae being injected into the host from the infected larva can be picked up by the previously uninfected larva. This acquired infection can then be passed through subsequent immature stages to adult mites, females of which will lay rickettsiae-infected eggs.

When trombiculid larvae feed on infected rodents in experimental conditions it is often difficult to infect them, and even if they do become infected the rickettsiae may not be transovarially transmitted to their progeny. It appears that larvae infected by feeding on humans can pass the infection to their own progeny by transovarial transmission, and subsequently to other humans and rodents. *Leptotrombidium* mites themselves are the main *reservoirs* of infection.

19.5   CONTROL

Repellents applied to the skin, such as DEET, dibutyl phthalate, dimethyl carbamate or benzyl benzoate can help reduce the likelihood of people becoming infested with mites. Clothing, especially socks, can also be impregnated with repellents or with permethrin.

The wide and patchy distribution of chigger mites make their control very difficult. However, if mite islands are identified then possibly scrub vegetation can be destroyed mechanically, with herbicides or by burning. This approach is not practical if mites are inhabiting cultivated land where ground vegetation consists mainly of crops. Spraying vegetation harbouring mites with fenthion, malathion, propoxur or permethrin can substantially reduce mite populations. Insecticides (acaricides) can be applied from knapsack sprayers, or from equipment mounted on vehicles or aircraft which generates insecticidal aerosols or fogs. Ultra-low-volume (ULV) spraying with permethrin or propoxur can also be undertaken.

FURTHER READING

Audy, J. R. (1968) *Red Mites and Typhus*. London: Athlone Press.

Azad, A. F. (1986) Mites of public health importance and their control. Mimeographed document WHO/VBC/86.931. Geneva: World Health Organization.

Frances, S. P., and Khlaimanee, N. (1996) Laboratory tests of arthropod repellents against *Leptotrombidium deliense* – noninfected and infected with *Rickettsia tsutsugamushi* – and noninfected *L. fletcheri* (Acari: Trombiculidae). *Journal of Medical Entomology*, **33**, 232–5.

Frances, S. P., Watcharapichat, D., Phulsuksombati, D. and Tanskul, P. (2000) Transmission of *Orientia tsutsugamushi*, the aetiological agent for scrub typhus, to co-feeding mites. *Parasitology*, **120**, 601–7.

Kawamura, A., Tanaka, H. and Tamura A. (eds.) (1995) *Tsutsugamushi Disease*. Tokyo: University of Tokyo Press.

Mount, G. A., Grothaus, R. H., Baldwin, K. F. and Haskings, J. R. (1975) ULV sprays of propoxur for control of *Trombicula alfreddugesi*. *Journal of Economic Entomology*, **68**, 761–2.

Roberts, S. H. and Zimmerman, J. H. (1980) Chigger mites: efficacy of control with two pyrethroids. *Journal of Economic Entomology*, **73**, 811–2.

Sasa, M. (1961) Biology of chiggers. *Annual Review of Entomology*, **6**, 221–44.

Strickman, D. (2001) Scrub typhus. In: *The Encyclopedia of Arthropod-Transmitted Infections of Man and Domesticated Animals*, ed. M. W. Service. Wallingford: CABI. pp. 456–62.

Takahashi, M., Murata, M., Misumi, H., Hori, E., Kawamura, A. and Tanaka, H. (1994) Failed vertical transmission of *Rickettsia tsutsugamushi* (Rickettsiales: Rickettsiaceae) acquired from rickettsemic mice by *Leptotrombidium pallidum* (Acari: Trombiculidae). *Journal of Medical Entomology*, **31**, 212–6.

Traub, R. and Wisseman, C. L. (1968) Ecological considerations in scrub typhus. 1. Emerging concepts. *Bulletin of the World Health Organization*, **39**, 209–18.

   (1968) Ecological considerations in scrub typhus. 2. Vector species. *Bulletin of the World Health Organization*, **39**, 219–30.

   (1968) Ecological considerations in scrub typhus. 3. Methods of area control. *Bulletin of the World Health Organization*, **39**, 231–7.

   (1974) The ecology of chigger-borne ricksettsiosis (scrub-typhus). *Journal of Medical Entomology*, **11**, 237–303.

20

Miscellaneous mites

Figure 20.1 Ventral view of hair follicle mite (*Demodex folliculorum*) showing four pairs of stumpy legs.

20.1 DEMODICIDAE

20.1.1 Follicle mites – *Demodex* species

Two species of *Demodex* commonly infect humans, namely *Demodex folliculorum* (Fig. 20.1) and *D. brevis*. The former is the more elongate species (0.2–0.4 mm long) and primarily inhabits the hair follicles, whereas *D. brevis* is squatter (0.1–0.2 mm) and lives in the sebaceous glands. Both species have a striated body and four pairs of very short stubby legs; they are remarkably non-mite-like.

Demodex mites, found in **hair follicles** or **sebaceous glands**, feed on subcutaneous tissues, especially sebum, and are particularly common on the forehead, nose, eyelids and cheeks adjacent to the nose. They have also been found on other parts of the body, in ear wax, and in the extruded contents of comedones ('blackheads'). Females lay eggs which hatch to produce six-legged **larvae** which moult to give rise to **protonymphs** then nymphs and finally adults. All the developmental stages, which extend over 13–15 days, occur within the hair follicles or sebaceous glands. Transfer of mites is believed to occur between mother and infants during close contact of nursing. Incidence of infection increases with age, and it seems that 90–100% of older people have these mites!

Usually people are unaware they have mites because they have not produced any adverse effects and control methods are therefore unnecessary. But occasionally mites cause dermatitis, such as acne, rosacea, impetigo contagiosa or blepharitis. Daily washing with a sulphur soap can reduce infections. In severe infections 'Danish ointment' which is a polysulphide preparation can be used, but this should not be applied to the eyelids as it can cause irritation. Alternative treatments consist of applying 0.5% selenium sulphide cream, 10% sulphur or 0.5% HCH cream (Kwell).

20.2 PYROGLYPHIDAE: HOUSE-DUST MITES

The family Pyroglyphidae contains five genera, but mites in the genus *Dermatophagoides* are the most important medically. *Dermatophagoides pteronyssinus* is known as the European house-dust mite and *D. farinae* as the American house-dust mite, but both species are found more or less

Figure 20.2 Ventral view of a house-dust mite (*Dermatophagoides pteronyssinus*).

worldwide. *Dermatophagoides* (Fig. 20.2) are very small mites (0.3 mm) that live among bedclothes, mattresses, carpets and general house dust. Female mites lay about 1–3 eggs a day. These hatch after 6–12 days and a six-legged *larva* emerges, which feeds and passes through two *nymphal* stages before becoming an adult. The complete life cycle takes about 3–4 weeks. *Beds* are the most important, and sometimes only, breeding site. Mites feed on fungi growing on floors and mattresses, discarded skin scales, semen and other organic debris. *High humidities* are required for mite survival and breeding. Adults live about 1–2 months. They are very rarely seen but become airborne when beds are made and are easily inhaled, together with their faeces, and this can cause allergic symptoms resulting in asthma and rhinitis.

Typically there may be 300 mites per gram of house dust. Under ideal breeding conditions some 5000 mites may be found in just 1 gram of mattress dust, but up to 15 600 has been recorded! *Densities* above 100 mites per gram are considered a risk factor for sensitization to allergies such as rhinitis, conjunctivitis, atopic eczema and asthma, while 500 mites per gram is a major risk factor in the development of acute asthma in those allergic to house-dust mites.

People suffering from severe allergies caused by house-dust mites can receive a series of injections with a specific desensitizing vaccine. Vacuum-cleaning carpets may have little effect on removing live mites because they cling firmly to carpet fibres, but may be more effective in removing mites from mattresses and sheets. Enclosing mattresses and pillows in plastic covers can help reduce mite infestations, although sleeping under such conditions may be uncomfortable. Washing bedding in water above 55 °C or dry-cleaning will kill mites, as will leaving electric blankets switched on at maximum temperature for 6 hours or more during the day. Treating sheets with 5% benzyl benzoate can be effective in killing mites, as can spraying carpets with pirimiphos methyl or permethrin. However, it may be difficult to reduce mite densities sufficiently to give any reduction in allergies suffered by householders.

20.3   OTHER MITES

There are numerous other mites which are parasitic on mammals and birds, including pets and livestock, and even on insects, and some of them occasionally become parasitic on people. For example, *Pyemotes tritici* (= *P. ventricosus*) usually parasitizes larvae of grain moths and beetles and is known as the grain mite or hay or straw itch mite, but it can bite people handling infested grain and straw. The cosmotropical rat mite (*Ornithonyssus bacoti*), the tropical fowl mite (*O. bursa*) and the chicken mite (*Dermanyssus gallinae*) may sometimes attack people, especially those working closely with infested animals. The bites from these mites can cause irritation and dermatitis. In addition, a few species such as *Liponyssoides sanguineus* which normally is ectoparasitic on mice and rats can spread rickettsial pox (*Rickettsia akari*), and this and a few other species can also transmit Q fever (*Coxiella burnetii*) to people.

There are also other mites, e.g. so-called forage mites that live among stored products such as grain, copra, flour, dried fruits, cheese and animal feeds. People habitually handling these substances may develop allergic symptoms such as dermatitis, and more rarely bronchitis and asthma. This may lead to terms such as 'grocer's itch', 'baker's itch', and 'copra itch' to describe these occupational hazards.

FURTHER READING

Azad, A. F. (1986) Mites of public health importance and their control. Mimeographed document WHO/VBC/86.931. Geneva: World Health Organization.

Cameron, M. M. (1997) Can house dust mite-triggered atopic dermatitis be alleviated using acaricides? *British Journal of Dermatology*, **137**, 1–8.

Colloff, M. J., Ayres, J., Carswell, F., Lowarths, P. H., Merrett, T. G., Mitchell, E. G., Walshaw, M. J., Warner, J. O., Warner, J. A. and Woodcock, A. A. (1992) The control of allergens of dust mites and domestic pets: a position paper. *Clinical and Experimental Allergy*, **22**, 1–28.

Fain, A., Guérin, B. and Hart, B. J. (1990) *Mites and Allergic Disease*. Varennes en Argonne, France: Allerbio.

Harvey, P. and May, R. (1990) Matrimony, mattresses and mites. *New Scientist*, 3 March, 48–9.

Krantz, G. W. (1978) *A Manual of Acarology*. 2nd edn. Corvallis, OR: Oregon State University.

Løvik, M., Gaarder, P. I. and Mehl, R. (eds.) (1998) The house-dust mite: its biology and role in allergy. Proceedings of an international scientific workshop, Oslo, Norway, 4–7 September 1997. *Allergy* **53** (Suppl. 48), 1–135.

McDaniel, B. (1979) *How to Know the Mites and Ticks*. Dubuque, IA: W. C. Brown.

Mumcuoglu, Y. (1976) House dust mites in Switzerland: I. Distribution and taxonomy. *Journal of Medical Entomology*, **13**, 361–73.

Nutting, W. B. (ed.) (1984) *Mammalian Diseases and Arachnids. Volume 1*: *Pathogen Biology and Clinical Management. Volume 2: Medico-Veterinary, Laboratory, and Wildlife Diseases, and Control*. Boca Raton, FL: CRC Press.

Owen, S., Morganstern, M., Hepworth, J. and Woodcock, A. (1990) Control of house dust mite antigen in bedding. *Lancet*, **335**, 396–7.

Wharton, G. W. (1976) House dust mites. *Journal of Medical Entomology*, **12**, 577–621.

Appendix   Names of some chemicals and microbials used in vector control (with common trade names in parentheses)

Carbamates
Bendiocarb (Ficam)
Carbaryl (Sevin)
Propoxur (Baygon, Arprocarb)

Chemosterilants
Apholate
Tepa
Thiotepa

Formamidines
Amitraz
Chlordimeform

Inorganics
Borax
Paris Green

Insect growth regulators (IGRs)
Cyromazine
Diflubenzuron (Dimilin)
Fenoxycarb
Hydrophen
Kinoprene
Lufenuron
Methoprene (Altosid)
Pyriproxyfen
Triflumuron

Microbials
Bacillus sphaericus (Vectolex)
Bacillus thuringiensis var. *israelensis* (*B.t.i.*) (= *Bacillus thuringiensis* H-14)
(Aquabac, Teknar, Vectobac)

Monomolecular films
Isostearyl alcohols
Lecithins

Natural organics
Azadirachtin (Neem)
Petroleum oils
Pyrethrum

Nicotinoids
Imidacloprid

Oganochlorines
DDT
Dieldrin
Endosulfan
Gamma-HCH (Lindane)
HCH (BHC)
Methoxychlor
Toxaphene (Camphechlor)

Organophosphates
Bromophos
Chloropyrifos (Dursban)
Coumaphos
Diazinon
Dichlorvos (DDVP, Vapona)
Dimethoate
Dioxathion (Hercules)
Fenchlorphos (Fenchlorvos, Ronnel)
Fenitrothion (Sumithion, Nuvanol)
Fenthion (Baytex)
Iodofenphos (Iodfenphos, Jodfenphos)
Malathion
Naled (Dibrom)
Pirimiphos methyl (Actellic)
Temephos (Abate)
Trichlorphon (Dipterex)

Pheromones
Muscalure

Pyrethroids
Allethrin
Alpha-cypermethrin
Bioallethrin
Cyfluthrin
Cypermethrin (Cymbush)
Deltamethrin (Decis)
Lambda-cyhalothrin (Icon)
Permethrin (Ambush)
Phenothrin (Sumithrin)

Repellents
Autan (Byrepel)
Benzyl benzoate (Ascabiol)
Citronella oil
DEET, i.e. diethyl-methylbenzamide (formerlly called diethyltoluamide)
Dibutyl phthalate
Dimethyl carbamate
DIMP, i.e. dimethylphthalate
Lemon eucalyptus oil

Synergists
Piperonyl butoxide

Glossary of common terms relevant to medical entomology

This glossary contains terms that are used in this book and a few others that are not, but which are considered pertinent to vector biology and control. In general morphological names are excluded as these are adequately explained in the preceding chapters. As usual with glossaries the inclusion and exclusion of terms is somewhat subjective.

Acaricide   Chemical that kills mites and ticks. Most acaricides are also insecticides.

Aestivation   Condition in which an organism has a period of inactivity as an adaptive response to unfavourable conditions encountered during the summer, or in the tropics the hot or dry seasons, e.g. aestivating adults of some *Anopheles* species.

Afrotropical region   (formerly Ethiopian zoogeographical region) The countries of Africa south of mid-Sahara, including southern Arabia, Madagascar, Seychelles and Cape Verde Islands. *See* Zoogeographical regions.

Amastigote   Morphological form of species of *Leishmania* and *Trypanosoma* with a rounded body and without a flagellum that occurs predominantly in macrophages (*Leishmania* species) or muscle cells (*T. cruzi*) of a vertebrate host.

Amplifying host   Usually used in relation to arboviruses where a host attains a very high level of parasites in its blood (high viraemic titre) which makes it very infective to vectors. For example, pigs develop high viraemias of Japanese encephalitis and are important amplifying hosts, as are monkeys in the transmission of Kyasanur Forest disease. Often not an efficient host for long-term maintenance of parasite populations.

Anautogenous   Refers to females of blood-sucking insects that require at least one blood-meal to develop their eggs. *See* Autogenous.

Anthropophagic   (anthropophilic) Refers to females of blood-sucking arthropods that prefer to feed on humans rather than other hosts. The degree of anthropophagy can vary geographically within species as well as between species. *See* Host preference.

Aperiodic   (non-periodic) Not exhibiting periodicity. When applied to microfilariae of helminths it means they are neither periodic nor subperiodic in their appearance in peripheral vertebrate blood, e.g. *Onchocerca volvulus*.

Arbovirus From *ar*thropod-*bo*rne *virus*. A virus that multiplies in a blood-sucking arthropod and is principally transmitted by the bite of arthropods to vertebrate hosts, e.g. yellow fever virus. Viruses, such as myxoma virus in rabbits, transmitted mechanically by arthropods (e.g. fleas and mosquitoes) and involving no cyclical development in the vector are not arboviruses.

Australasian region Australia, Tasmania, New Zealand, Melanesia, Micronesia and Polynesia. *See* Zoogeographical regions.

Autogenous Refers to females of blood-sucking insects (usually mosquitoes and simuliids) that lay at least the first batch of eggs without a blood-meal. Subsequently a blood-meal is required for each further oviposition. Some mosquitoes, e.g. *Toxorhynchites* species, never blood-feed and are thus always autogenous. *See* Anautogenous.

Biodegradable insecticide Insecticide that does not accumulate in the environment but is broken down by micro-organisms in the soil and water into relatively harmless compounds, e.g. the organophosphates and carbamates.

Biological control (biocontrol) Deliberate introduction, or augmentation, of biological agents such as pathogens, parasites and predators (especially fish) to control arthropod populations, mainly mosquito larvae. Although *Bacillus thuringiensis* var. *israelensis* is often included in this category it is not a biological control agent because it does not recycle and persist in the environment but is applied as a non-living formulation and is thus more accurately described as a microbial insecticide.

Biological transmission Transmission of disease organisms with biological involvement between the vector and parasite. Can involve (1) multiplication without change in form, e.g. *Yersinia pestis* in fleas, arboviruses in ticks and mosquitoes; (2) multiplication with change in form, e.g. *Plasmodium* species in *Anopheles*; or (3) no multiplication (usually a decrease in parasite numbers) but with change in form, e.g. filarial parasites in mosquitoes and black-flies.

Borrow pit Excavation, often by the side of a road or railway, dug to provide earth for buildings or embankments. These fill with water and become mosquito larval habitats.

Bromeliad An epiphytic plant growing on trees in the Americas. Its water-filled leaf axils are colonized by mosquito larvae. *See* Epiphyte.

Campestral In epidemiology, used to describe transmission occurring in fields and open spaces, such as plague transmission among wild rodents. But also used sometimes to include transmission in woods and forests, which strictly is sylvatic transmission.

Carbamates Synthetic insecticides which are derivatives of carbamic acid, e.g. carbaryl and propoxur.

Chemosterilant Chemical used to induce sterility, but not usually death, in arthropods so as to control them, e.g. apholate and tepa. Chemosterilized insects are sometimes used in genetic control of insect vectors.

Chitin A constituent of arthropod cuticle giving it a hard or leathery texture. Some insect growth regulators (IGRs) prevent chitin formation and are used in the control of vectors.

Co-feeding Sometimes an infection can be passed from an infected vector feeding on a host to a nearby uninfected vector. This phenomenon has been recorded in ticks and mites, such as in the transmission of scrub typhus.

Commensal Animals living together or in close association, e.g. roof rats (*Rattus rattus*) living in and around human habitations.

Cytoform (cytotype) A cytologically defined population with a distinctive chromosomal complement, that may be a species or considered as a chromosomal variant within a species. In the latter case individuals are often given vernacular geographical names, e.g. *Anopheles gambiae* Savanna and Forest forms. Different cytoforms may exhibit differences in behaviour.

Dead-end host Host that although infected with an arbovirus, or other pathogen, and may be severely affected, has a viraemia too low in titre to infect blood-feeding arthropods. Examples are humans in the epidemiology of Japanese encephalitis, and horses in western equine encephalitis transmission.

Definitive host Host in which parasites reach maturity. This rarely occurs in arthropod vectors, but the noted exception is the development of malarial parasites, involving sexual reproduction, in mosquitoes. *See* Intermediate host.

Diapause State of inactivity or arrested development which allows an organism to survive a period of unfavourable conditions. Development cannot be resumed, even under favourable conditions, until diapause is 'broken' by environmental stimuli such as changes in photoperiod (length of daylight) or temperature.

Dichoptic A condition of the head in adult Diptera in which the eyes are widely separated from each other, as, for example, in female adult simuliids. *See* Holoptic.

Diel periodicity Periodicity that occurs about every 24 hours.

Diurnal Refers to activity during the daylight hours, such as blood-feeding in simuliids and appearance in peripheral vertebrate blood of microfilariae of some helminth species, e.g. *Loa loa*.

Emulsion Suspension of minuscule droplets of one liquid in another. For example, when an insecticide dissolved in oil is mixed with water this results in a milky emulsion.

Endemic Describes a disease in a human population that is constantly present and more or less stable. The numbers of new cases cannot exceed the numbers of new susceptible individuals entering the population, but may be fewer. The opposite is epidemic. In animal populations the equivalents are enzootic and epizootic.

Endophagic Describes insects, such as some mosquitoes, that enter houses to blood-feed.

Endophilic Describes insects, such as some mosquitoes, that rest in houses before or after blood-feeding in houses or outside.

Endotoxin Toxin formed inside bodies in Gram-negative bacteria which is released only when the bacteria are lysed (broken down). An example is the toxins in the parasporal body of *Bacillus thuringiensis* var. *israelensis*, which on ingestion are lethal to simuliid and mosquito larvae.

Epidemic Occurrence of a disease in the human population where the numbers of cases exceed the normal expected number of cases. An epidemic situation can be only temporary. *See* Endemic.

Epimastigote (crithidial form) Morphological form of a trypanosome with the flagellum emerging about halfway in the body but remaining attached to the cell membrane, e.g. trypanosomes in the salivary glands of tsetse-flies and mid-gut of triatomine bugs. This is not the stage that is infective to the vertebrate host. *See* Metacyclic trypanosome.

Epiphyte Plant that grows on other plants, usually trees, but without being parasitic. It derives its water mainly from rain. Bromeliads, the water-filled leaf axils of which provide mosquito larval habitats, are epiphytes.

Eurasia A geographical area comprising the continental land mass of Europe and Asia combined.

Exophagic Term applied to insects that blood-feed outside houses, e.g. *Aedes aegypti* and simuliid species.

Exophilic Term applied to blood-sucking insects that rest outside houses, irrespective of whether they have fed inside or outside houses.

Exoskeleton Outer body layer, often called the integument or cuticle, of arthropods that is shed during moulting from one life-stage to another, e.g. third-instar mosquito larva to fourth-instar, and then fourth-instar larva to pupa.

Extrinsic incubation period Duration of the part of a parasite's life cycle that is completed in a vector, that is the time from a vector becoming infected to it being infective (i.e. capable of transmitting the parasite).

Furuncular myiasis Dipterous larvae living in boil-like swellings, e.g. larvae of the tumbu fly.

Genetic control Special type of biological control that uses genetical techniques to control pest populations. Specifically the reduction of the reproductive potential of vectors. For example, the release of sterilized male vectors into field populations which results in producing large numbers of mated, but infertile, females. These females lay eggs that cannot hatch, so causing a population decrease. *See* Sterile male release.

Gonotrophic cycle   Time from first blood-feeding to oviposition, and subsequently between successive ovipositions. The first such gonotrophic cycle may be a day or two longer than subsequent ones. Sometimes the gonotrophic cycle is defined as time from blood-feeding to blood-feeding. This cycle is sometimes called the ovarian cycle.

Habitat   Usually means the physical environment in which an animal lives, e.g. the skin in the case of scabies mites, streams for simuliid larvae and animal nests for many ixodid ticks.

Haemocoel   Main body cavity of arthropods in which insect blood (haemolymph) circulates.

Haemolymph   Insect blood, usually colourless and clear, which fills the haemocoel.

Hemelytron   (pl. hemelytra) Fore-wing of certain insects, e.g. triatomine bugs, which has a thickened basal part and a membranous distal part.

Hemimetabolous   (incomplete metamorphosis) Describes the development from egg to adult which is gradual, passing through one or more nymphal stages, e.g. lice and ticks. If adults are winged the nymphal wing buds grow larger at each moult, e.g. bedbugs. There is no pupal stage. *See* Holometabolous.

Hereditary transmission   Involves a female vector passing disease organisms to her eggs and thus to the next generation, i.e. transovarial transmission.

Hibernation   Period of inactivity and/or altered behaviour caused by cold conditions, e.g. winter. Some mosquitoes, e.g. *Culex pipiens* in Europe, may remain in complete hibernation without blood-feeding for many months by living off their fat reserves. Other species, e.g. *Anopheles atroparvus*, enter incomplete hibernation during which time adults need to emerge periodically from hibernation sites to take blood-meals to renew their fat reserves.

Holarctic   Sometimes used to encompass the Palaearctic and Nearctic regions. *See* Zoogeographical regions.

Holometabolous   (complete metamorphosis) Arthropod development from egg to adult in which the body form changes completely in appearance. A pupal or puparial stage is characteristic of holometabolous development. *See* Hemimetabolous.

Holoptic   A condition of the head of adult Diptera in which the eyes meet or nearly meet each other, as, for example, in male adult simuliids. *See* Dichoptic.

Host preference   Preferred hosts (e.g. species, sex, age) of an arthropod in an area when a choice exists. For a particular species this can alter from area to area as well as seasonally, depending usually on the availability of alternative hosts.

IGRs   (insect growth regulators) Sometimes known as insect development inhibitors, these are groups of chemicals that either prevent the development of larvae into pupae or pupae into adults (juvenile hormone analogues, e.g. methoprene) or interfere with the moulting process, killing larvae as they moult (chitin synthesis inhibitors, e.g. difluben-zuron).

Impoundment Well-defined man-made area of water, usually with more-or-less vertical sides, often constructed to provide water for domestic, agricultural or recreational puposes. However, impound-ments are dug sometimes principally to reduce mosquito breeding in formerly marshy areas.

Infected Applied to arthropods when a parasitic infection has been taken up by the vector but is not yet in a stage in which it can be transmitted to a host. Examples are simuliid adults with onchocercal larvae in their flight muscles, or a vector with arboviruses in the stomach but not yet in the salivary glands. *See* Infective.

Infective Applied to arthropods when the parasites can be transmitted by the vector to a host. Examples are mosquitoes with malarial sporozoites in the salivary glands or simuliids with third-stage onchocercal worms in their mouthparts. *See* Infected.

Insecticide resistance Ability of arthropods to tolerate doses of insecti-cide which would prove lethal to the majority of normal (susceptible) individuals of the same species. Rare mutants which are resistant are selected for by the use of the insecticide.

Instar One of a series of life-cycle stages in metamorphosis that are sepa-rated by a moult, e.g. the first, second and third larval instars of house-flies and the five nymphal stages of bedbugs.

Integument The cellular epidermis and outer non-cellular cuticle which together provide the outer covering of arthropods. *See* Exoskeleton.

Intermediate host Host in which a parasite does not reach sexual matu-rity. Applies to most parasites in arthropod vectors, e.g. filarial para-sites in mosquitoes and simuliids. *See* Definitive host.

Intrinsic incubation period Duration of the life cycle of a parasite in the vertebrate host; interval between infection and first clinical symptoms.

Karyotype The number and appearance of the chromosomes in the nuclei of a species.

Larviparous Reproduction in which the egg(s) hatch within the female and larva(e) are deposited, e.g. tsetse-flies.

Life cycle (life history) In entomology and parasitology this usually means the series of morphological stages an organism passes through to reach the mature adult stage, and the biology of each stage.

Longevity How long an organism lives, often expressed as the mean expectancy of life. Vector longevity is one of the most important factors in disease transmission dynamics and vector control.

Maggot Legless larva that has no distinct head, thorax or abdomen, e.g. larvae of house-flies.

Maintenance host A vertebrate or arthropod host which allows long-term survival of parasite populations. The host must have an infection rate that is at least adequate to maintain a population of the disease agent continuously endemic in an area. Humans can be maintenance hosts of louse-borne typhus, and mosquitoes and birds appear to be maintenance hosts of some of the encephalitis viruses. A maintenance host may retain an infection during periods when there is no or little vector transmission, e.g. during dry seasons or winters.

Mechanical transmission   Transmission where there is no multiplication or cyclical development of the aetiological agent (i.e. parasite or pathogen), it being merely passively carried by the vector. Examples are house-flies transmitting trachoma virus and dysenteries by their feet, vomit or faeces, and trypanosomes transmitted by stable-flies (*Stomoxys* species).

Metacyclic trypanosome   (metatrypanosome) The final, and usually smaller, version of the trypomastigote form in the vector that is infective for the vertebrate host.

Metamorphosis   Changes in form from the first stage (egg) in the life cycle of an arthropod to the adult form. In hemimetabolous arthropods, e.g. bedbugs, lice, ticks and mites, the change is gradual through nymphal stages which resemble small versions of the adult. In holometabolous arthropods, e.g. mosquitoes, tsetse-flies and fleas, the changes are abrupt and involve larval stage(s) and a pupal or puparial stage which are very dissimilar to the adult.

Microbial insecticide   Insecticide comprising a biological agent such as bacteria, e.g. *Bacillus thuringiensis* var. *israelensis* and *B. sphaericus*, or toxic compounds derived from such agents.

Middle East   Many definitions but here covers land surrounding southern and eastern shores of the Mediterranean Sea, extending from Moroco to the Arabian Peninsula, including Egypt, Turkey, Iran, Iraq, Sudan and Libya.

Monolayer   Thin film of a water-insoluble surfactant, e.g. lecithins, that has a very high spreading power on water. Mosquito larvae are unable to maintain normal contact with the water surface, and this combined with the reduction of dissolved oxygen content caused by monolayers causes their death.

Morphology   The outward structure of an organism. Most arthropods are identified by their morphology, that is by their outer appearance.

Moulting   Process of shedding the cuticle between developmental stages (i.e. instars).

Myiasis   Invasion of vertebrate organs or tissues by larvae of Diptera that feed on living or dead tissues.

Nearctic region   The USA, Canada, Greenland and northern Mexico. *See* Zoogeographical regions.

Neotropical region   South America, Central America, southern Mexico and Caribbean islands. *See* Zoogeographical regions.

New World   North, Central and South America, and the Caribbean area. *See* Old World.

Nidicolous   In medical entomology usually used to describe the habit of soft, and some hard, ticks of living in and around the homes, nests, burrows, and caves of their hosts. Such ticks disperse little. *See* non-nidicolous.

Nocturnal   Refers to activity during the night, such as blood-feeding in anophelines and appearance in vertebrate blood of nocturnally periodic microfilariae of some helminths, e.g. *Wuchereria bancrofti.*

Non-nidicolous   Describes the habits of most hard ticks of living in open and exposed habitats away from their hosts' homes. Such ticks are often dispersed by mammals and birds over considerable distances. *See* nidicolous.

Nymph   In incomplete metamorphosis (hemimetabolous development) the stage in the life cycle that hatches from the egg (e.g. bedbugs, lice), or the stage that arises through the moulting of the larval stage (e.g. ticks).

Old World   All countries and areas east of the Americas. *See* New World.

Oocyst rate   Percentage of mosquitoes that have malarial oocysts on the stomach.

Organochlorines   (chlorinated hydrocarbons) Synthetic insecticides containing carbon, chlorine and hydrogen, e.g. DDT, dieldrin, HCH and methoxychlor.

Organophosphates   Synthetic insecticides which are derivatives of phosphoric acid and hence all contain phosphorus, e.g. diazinon, dichlorvos and malathion.

Oriental region   Asia east of Pakistan and south of the Himalayas and central China, covering Taiwan, Sri Lanka, and the South-east Asia archipelago eastwards to include Sulawesi. *See* Zoogeographical regions.

Ornithophagic   (ornithophilic) Arthropods that blood-feed on birds.

Osmoregulation   The regulation of water balance in arthropods; maintaining the homeostasis (balance) of osmotic and ionic content of the body fluids.

Overwintering   Describes the survival tactics of arthropods during winters. For instance adults (e.g. some mosquitoes) may cease feeding and ovipositing and enter a state of hibernation until warmer weather reappears and activity resumes. The growth and development of the immature stages (e.g. mosquito and simuliid larvae) may also slow down.

Palaearctic region   Europe, North Africa, Asia north of the Himalayas and central China, Japan, Iceland, mid-Atlantic islands. *See* Zoogeographical regions.

Peritrophic membrane   A thin tubular sheath secreted either by cells at the anterior end of the mid-gut (mosquito larvae, tsetse-fly adults) or by cells lining the mid-gut (adult mosquitoes). It is found only in some insects. It supposedly forms a protective lining for the mid-gut, but its exact function remains largely unknown. In tsetse-flies the peritrophic membrane plays a role in the cyclical development of human sleeping sickness trypanosomes.

Pheromone   Chemical (semiochemical) released usually as an odour by an individual which produces reactions in others of the same species, e.g. sex pheromones in tsetse-flies and ticks.

Phoresy   Transport of an animal from one place to another by means of attachment to another animal, e.g. *Simulium neavei* larvae on freshwater crabs, and eggs of *Dermatobia* flies attached to adult female mosquitoes.

Polynesia   A group of numerous islands in the west Pacific extending from Hawaii to New Zealand and including the Solomon Islands, New Caledonia, Fiji, Tuvatu, Tonga, Samoa, Cook Islands, Society Islands, and Tahiti.

Polytene chromosomes   So-called giant chromosomes found only in certain tissues of Diptera, such as the ovarian nurse cells of half-gravid anophelines and larval salivary glands of simuliids. When stained they show distinct banding patterns which can often be used to identify species within species complexes. *See* Sibling species.

Promastigote   (leptomonad) Morphological form of a trypanosomatid with the flagellum arising near the anterior end, e.g. *Leishmania* parasites in the phlebotomine sand-fly gut.

Pseudopod   (false leg) Stumpy protuberances present on dipterous larvae of some species (e.g. tabanids, phlebotomine sand-flies) that assist them in locomotion. None of the larvae of the Diptera possesses true legs.

Pseudotracheae   Small tubes in the labella of the adults of some Diptera (e.g. muscids, calliphorids and tabanids) which are supported by sclerotized rings. Liquid food passes through minute openings in these pseudotracheae to the mouth of the fly.

Puparium   Life-stage resulting from the hardening and sclerotization of the cuticle of the last larval instar of certain Diptera, e.g. tsetse-flies and house-flies. Equivalent to the pupa of other insects.

Pyrethroids   (synthetic pyrethroids) Synthetic insecticides containing different pyrethrin-like chemicals, e.g. permethrin, deltamethrin, lambda-cyhalothrin.

Questing   The behaviour of ticks, mainly ixodids, when climbing up vegetation, such as grasses and herbaceous plants, in order to actively seek out passing hosts, to which they attach themselves.

Quiescence   Temporary state of arrested or slowed development, such as in ixodid larvae after blood-feeding but prior to moulting to the nymphal stage, or state of inactivity of some hibernating adult mosquitoes.

Reservoir hosts   Animals in which populations of disease organisms persist indefinitely, and which pass the disease to other species of hosts, often by vectors. Reservoir hosts may be maintenance hosts, and are often mammals and birds. Some arthropod vectors which are long-lived, and may also be capable of transovarial transmission, are sometimes regarded as reservoir hosts, e.g. ticks as vectors of relapsing fever and various viruses.

Residual spraying Application of a persistent insecticide (e.g. malathion) to surfaces, such as to the inside walls and roofs of houses in malaria control programmes, and to trees in tsetse control projects.

Rickettsiae A group of Gram-negative intracellular coccoid-shaped bacteria, many of which are transmitted by arthropods, e.g. *Rickettsia prowazekii* transmitted by body lice and causing typhus. Formerly rickettsiae were regarded as micro-organisms intermediate between bacteria and viruses.

Seed tick Name often given to the very small larva of an ixodid tick before it has blood-fed.

Sclerotization Process which results in the new arthropod cuticle formed after moulting being tanned (darkened) and hardened to give it rigidity.

Sibling species (isomorphic or cryptic species) Species that are morphologically indistinguishable or almost so, and which are recognized by non-morphological features, usually their polytene chromosomes (e.g. species of the *Simulium damnosum* complex). In nature they are reproductively isolated from each other. Sibling species may differ only slightly biologically but be significantly different in epidemiological importance.

s. l. (sensu lato) Means in the broad sense. When placed after the name of a species denotes that reference is made not just to that species but also to closely related species (e.g. sibling species) within a complex. *See* s. str.

Source reduction Simple measures that either prevent breeding of arthropods or eliminate their breeding sites. Mainly applicable to mosquito control, e.g. covering water tanks, filling in puddles or removing discarded water-retaining receptacles.

Species A group of individuals in natural populations that can interbreed by mating within the group and producing fertile progeny; individuals are usually similar in appearance and behaviour.

Species complex A group of sibling, or very closely related, species that are morphologically indistinguishable (isomorphic) but are reproductively isolated, and which often live in the same area (sympatric). Species within a complex can often be identified by their polytene chromosomes or by biochemical or molecular techniques. The Diptera contain many complexes, such as in mosquitoes (e.g. *Anopheles gambiae* complex) and simuliids (e.g. *Simulium damnosum* complex).

Species group Used for an assemblage of closely related species that although they may be morphologically similar in one or more life-stages can nevertheless be distinguished on external appearance as distinct species, e.g. the *Simulium neavei* group – comprising eight species including *S. neavei*, *S. woodi* and *S. nyasalandicum*.

Species sanitation Control, or eradication, directed against just one species of vector or pest in a particular area, usually using simple techniques.

Spermatheca One or more receptacles in the female reproductive system of arthropods which receive sperm during mating.

Spirochaetes Gram-negative bacteria which have a more-or-less spiral shape, e.g. *Borrelia duttonii* which is spread by soft ticks and causes relapsing fever.

Sporogony That part of the sexual cycle of sporozoans (e.g. *Plasmodium* species) in which sporozoites are produced.

Sporozoite rate Percentage of mosquitoes with malarial sporozoites in their salivary glands.

s. str. (sensu stricto) Means in the strict, or narrow, sense. When placed after a species name denotes that only that species is being referred to, and not any of its closely related (e.g. sibling) species in a complex. Sometimes abbreviated to s. s. *See* s. l.

Sterile male release (sterile-insect technique – SIT) In genetic control programmes the inundative release of large numbers of artificially sterilized male arthropods into field populations in the hope that this will result in sterile matings and consequently a reduction in population size.

Subperiodic As applied to microfilariae in peripheral vertebrate blood, means they exhibit partial diel (24-hour) periodicity. That is, their concentration in the blood decreases from a maximum to a minimum, but not close to zero as with microfilariae showing periodic periodicity. Diurnal subperiodicity occurs in the Pacific strain of *Wuchereria bancrofti* while nocturnal subperiodicity is found in populations of *Brugia malayi* in West Malaysia, Thailand, etc.

Subspecies In zoology the only taxon recognized below the rank of species. Subspecies differ morphologically from other members of the species and are spatially isolated, e.g. either geographically or by their hosts. Consequently they are normally reproductively isolated, but when brought together can interbreed with other subspecies. Subspecies may eventually evolve into distinct species.

Sylvatic In epidemiology means that diseases are contracted in woods or forests, e.g. the forest cycle of yellow fever. *See* Campestral.

Synanthropic Applied to animals living in close association with humans or their houses, e.g. house-flies and triatomine bugs.

Synergist Chemical that has little or no toxicity but, when combined with some insecticides, enhances their activity and thus reduces dosage rates. For example, piperonyl butoxide is a synergist added to the insecticide pyrethrum.

Transovarial transmission Production by an infected vector of eggs infected with parasites (e.g. viruses, rickettsiae). When they hatch they give rise to individuals (e.g. larvae, nymphs) that are infected and are either capable of transmitting the parasites (e.g. ticks) or passing it on to later life-cycle stages that transmit the infection (e.g. scrub typhus mites). *See* Hereditary transmission.

Transstadial transmission   Survival of parasites through successive life cycle stages (larva → nymph(s) → adult) of an organism (e.g. ticks), each of which can transmit the parasite if it is haematophagous.

Traumatic myiasis   The infestation of wounds by dipterous larvae, e.g. screwworm larvae. *See* Furuncular myiasis.

Trypomastigote   Morphological form of a trypanosome with the flagellum arising near the posterior end, and running the length of the body where it is attached to the cell membrane. Trypomastigotes are found in the vertebrate blood of hosts infected with trypanosomiasis, and are the form ingested by a vector with its blood-meal.

Ultra-low-volume   (ULV) Refers to spraying an insecticide in concentrated form; thus the dosage rate is small, e.g. <5 litres/hectare.

Vector   Organism that conveys an aetiological agent from one host to another. A vector may be an intermediate host (e.g. culicines transmitting filariasis) or not (e.g. house-flies mechanically transmitting bacteria).

Venereal transmission   When pathogens, such as viruses and rickettsia, are passed from congenitally infected males to females during mating, Occurs in some vectors of yellow fever and Crimean–Congo haemorrhagic fever.

Viraemia   Presence of virus in vertebrate blood. High viraemias are usually necessary for transmission by arthropod vectors.

Wettable powder   (water-dispersible powder) Technical grade insecticide diluted with an inert carrier (dust) and to which a wetting agent or surfactant has been added. The resultant wettable powder is then mixed with water for spraying onto surfaces.

Zoogeographical regions   Any one of the six main geographical areas referred to by zoologists. Each region has its own particular fauna, of which many species occur only in that region. The Palaearctic and Nearctic region are sometimes grouped together and referred to as the Holarctic region. *See* Afrotropical, Australasian, Nearctic, Neotropical, Oriental and Palaearctic regions.

Zoonosis   Natural transmission of infections between vertebrate hosts and humans.

Zoophagic   (zoophilic) Blood-sucking arthropods that feed on non-human animals.

Select bibliography

Beaty, B. J. and Marquardt, W. C. (eds.) (1996) *The Biology of Disease Vectors*. Boulder, CO: University Press of Colorado.

Chavasse, D. C. and Yap, H. H. (eds.) (1997) *Chemical Methods for the Control of Vectors and Pests of Public Health Importance*. WHO/CTD/WHOPES/97.2. Geneva: World Health Organization.

Clark, G. G. (coordinator) (1994) Prevention of tropical diseases: status of new and emerging vector control strategies. Proceedings of a symposium on vector control. *American Journal of Tropical Medicine and Hygiene*, **50** (6) (suppl.), 1–159.

Eldridge, B. F. and Edman, J. D. (eds.) (2000) *Medical Entomology: a Textbook on Public Health and Veterinary Problems Caused by Arthropods*. Dordrecht: Kluwer.

Kettle, D. S. (1995) *Medical and Veterinary Entomology*. 2nd edn. Wallingford: CAB International.

Lane, R. P. and Crosskey, R. W. (eds.) (1993) *Medical Insects and Arachnids*. London: Chapman and Hall.

Mullen, G. and Durden, L. (eds.) (2002) *Medical and Veterinary Entomology*. Amsterdam: Academic Press.

Service, M. W. (ed.) (2001) *The Encyclopedia of Arthropod-Transmitted Infections of Man and Domesticated Animals*. Wallingford: CABI.

van Emden, H. F. and Service, M. W. (2004) *Pest and Vector Control*. Cambridge: Cambridge University Press.

World Health Organization (1997) *Vector Control: Methods for Use by Individuals and Communities*, prepared by J. A. Rozendaal. Geneva: World Health Organization.

Index